AVOIDING CORROSION IN DESALINATION PLANTS

Roger Francis

NACE INTERNATIONAL

NACE International
The Worldwide Corrosion Authority

Printed in the United States of America

ISBN: 978-1-57590-390-3

NACE International
The Worldwide Corrosion Authority
15835 Park Ten Place
Houston, TX 77084
nace.org

Dedication

To all my friends and colleagues at the former Weir Group for sharing your knowledge and expertise with me over the years.

Table of Contents

Preface

I first became involved with corrosion problems in desalination plants soon after joining Weir Materials and Foundries in 1991, when I was asked to visit an MSF plant in the Middle East to investigate some heat exchanger failures soon after start up. In the 27 years since then, I have investigated failures in MSF, MED, and SWRO desalination plants involving a wide range of materials, including carbon steel, copper alloys, and stainless steels. Since I set up my own consulting company in 2012, desalination has been the industry I have been involved with the most. A lot of the failures and materials selection issues are not difficult, and corrosion problems in desalination plants have been reported in various conferences and journals, but I realized that there is no one place where data on corrosion in desalination plants is collected together. Hence, I conceived the idea of this book as a place to go for information on the common corrosion problems and how to prevent them. This book is intended for engineers and it does not specifically address the science behind desalination, except in so far as it is necessary to understand the corrosion issues.

I am indebted to a number of friends for sharing their data with me, and reviewing early drafts, in particular Charles Desportes, Carol Powell, Glenn Byrne, Geoff Warburton, Steve Clarke, and Stan Hebdon. I would like to thank Rolled Alloys and Intertek for the use of some of their photographs and technical information. I would also like to thank my brother Ian Francis, who did the original line drawings.

I hope that you find this document useful.

ROGER FRANCIS
February 2019

1 Introduction

The world's population is steadily increasing and with it is an increasing demand for water—for both drinking and irrigation. In many areas of the world, particularly in warmer climates, there are limited sources from rivers and wells, so desalination is being increasingly used to produce water to satisfy both requirements. Although desalination is sometimes carried out on brackish waters and highly saline well waters, most desalination plants generate fresh water from seawater.

There are three main processes used in desalination plants, the oldest of which is multistage flash (MSF), where the water is essentially boiled at low pressure and the steam that flashes off is condensed for drinking water. The second process is multiple-effect distillation (MED), in which low pressure steam is used to force evaporation of seawater and the vapor is then condensed for drinking water. Although actual MSF and MED plants (large-scale) are land based, small-scale units have been fitted to large ships, such as cruise liners, to generate fresh water.

The third process is seawater reverse osmosis (SWRO), where chloride is selectively removed from water by forcing it at high pressure through a special membrane. This method involves no heat transfer but requires enough electricity to power the high pressure pumps that are required. All three of these methods have advantages and disadvantages.

MSF plants require a large quantity of steam and are usually associated with a power station, which can generate extra steam and electricity for the desalination plant. They can produce very large quantities of fresh water and have been widely used in the Middle East, where they are associated with oil-fired power stations. They use relatively large amounts of energy to generate fresh water and may not be competitive with MED or SWRO in some applications.

MED plants were traditionally smaller than MSF plants, but only require a supply of low pressure steam. With improvements in scale control chemicals, large plants, of a similar size to MSF have become possible with a lower water production cost.

SWRO plants require no heat, but they do require significant quantities of electricity to drive all the high- and low pressure pumps. Some recent plants have constructed large solar arrays in the desert or have installed wind farms, so that they are powered by renewable energy. They do have the advantage that they can be constructed in a wide range of sizes (see Chapter 5), and can be modularized, so that extra modules–or "trains"–can be added as water demand increases. They are also popular in holiday destinations where water demand varies through the year, because SWRO is relatively easy to turn on and off. Most of the new desalination plants currently being designed and built are SWRO.

The energy requirements for each process are shown in Table 1.1. Note that reverse osmosis currently has the lowest energy requirements,[1] but this does not take into account the energy required to construct or refurbish items consumed in the process. However, the cost of the final fresh water is affected by a variety of factors, including capacity and type of the facility, location, feed water, labor costs, energy costs, financing arrangements, and cost of the brine disposal.

TABLE 1.1 Energy consumption of seawater desalination methods.[1]

DESALINATION METHOD	ENERGY CONSUMPTION (kWh/m^3)		
	MSF	MED	SWRO
Electrical Energy	4-6	1.5-2.5	3-5.5
Thermal Energy	50-110	60-110	None
Electrical Equivalent of Thermal Energy*	9.5-19.5	5-8.5	None
Total Equivalent Electrical Energy	13.5-25.5	6.5-11	3-5.5

* The amount of electrical energy that could be generated using a given quantity of thermal energy and a suitable turbine generator.

All three processes can be subject to corrosion unless suitable materials are used for the many components, and they are operated within their normal limits of use. Some of these corrosion mechanisms are common to two or all three types of desalination, while others are specific to one method.

In Chapter 2 the various metal alloys used in desalination plants are discussed, including their mechanical properties and any special problems. The properties of the common polymers used in desalination plants are also discussed, as well as the importance of selecting the correct grade of polymer appropriate for use with different aggressive fluids. Chapter 3 addresses welding of the metals and fabrication of the polymers, highlighting particular problems in fabrication and ways of avoiding them.

In Chapter 4, the corrosion resistance of various alloys is discussed. This is arranged by metal type, e.g., copper alloys, so that only the corrosion mechanisms affecting that type of alloy in a desalination environment are discussed. Methods of avoiding the specific corrosion types for each class are discussed.

Chapter 5 deals with SWRO plants, including how they work and the conditions in each part of the plant. The commonly used materials for the major items in each section of the plant are discussed. There is also a chapter section with case histories of recent failures, highlighting some of the statements on material selection and operation made earlier in the chapter. Following this is a section on possible future developments in this technology and the implications for corrosion.

In Chapter 6, the operation and working conditions of a typical MSF plant are discussed, along with the typical materials used in the various parts. The results of two surveys on the performance of materials in MSF plants are also discussed. There is also a section with case histories of recent failures, highlighting some of the statements on material selection and operation made earlier in this chapter. Following this is a section on possible future developments in this technology and the implications for corrosion.

In Chapter 7, MED plants and how they work and the conditions in each part of the plant are described. The commonly used materials for each of the major components in each section are discussed. There is a section on recent failures in MED plants, highlighting the materials selection statements made earlier in the chapter. Finally, the effects of possible future developments on corrosion are discussed.

Chapter 8 addresses standards, procurement, testing, and quality assurance/quality control (QA/QC), and highlights the need for additional requirements over and above those of ASTM standards if materials of adequate quality are to be procured. Some recent failures due to poor quality material are discussed, highlighting this need.

The alloys as listed in Unified Numbering System (UNS) standards[2] are used throughout, although they are usually referred to by their generic names; ASTM standards (actual ones are referenced in their respective chapters) are used to discuss the various product forms and test methods. The equivalent grades to other standards, such as European Norms (EN), can be found on several comparative websites on the Internet. Appendix A contains the nominal composition and generic names of all the cast- and wrought alloys discussed in this book so an alloy composition can be readily checked when reading any of the chapters. Appendix B contains details about how to do ferrous sulfate dosing when it is required to control corrosion of copper alloy heat exchanger tubes.

If this document is read from beginning to end, it will be noticed that there is a certain amount of repetition. This is because it is expected that most readers will dip into sections relevant to their current question and they will not wish to keep following references to other chapters. When the point at issue requires more lengthy discussion, references to other chapters has been unavoidable.

REFERENCES

1. "Energy Requirements of Desalination Processes," *Encyclopedia of Desalination and Water Resources* (DESWARE) (Abu Dhabi, UAE: Encyclopedia of Life Support Services), available from http://www.desware.net/Energy-Requirements-Desalination-Processes.aspx (February 1, 2018).
2. "Metals and Alloys in the Unified Numbering System" (Warrendale, PA: SAE International; and West Conshohocken, PA: ASTM International, 2004).

2 Materials

The compositions and properties of the materials commonly used in desalination plants are discussed in this chapter. The nominal compositions of the different alloys are collected in Appendix A, for ready reference from anyplace in the book.

2.1 Austenitic Cast Iron

Austenitic cast irons are similar to gray cast irons, but they have a substantial nickel addition to stabilize the face center cubic (FCC) austenite structure, rather than the body center cubic (BCC) ferrite structure of gray cast irons. The austenitic structure means that these alloys have increased ductility. The austenitic cast irons are covered by ASTM A439[1] and the compositions are shown in Table A.1. These are the D grades, which contain a small magnesium addition to spheroidize the graphite, which increases ductility. In addition to the nickel, most of the austenitic cast irons contain some chromium, which increases the strength by forming chromium carbides.

For seawater service, the D-2 grade is commonly used, although the D-2B grade has also been used because of its slightly better resistance to erosion. Table 2.1 shows the minimum mechanical properties of the austenitic cast irons in Table A.1. It can be seen that the ductile austenitic cast irons have moderate strength with some ductility and they have mostly been used for large seawater pump cases and column pipes. The use of the alloy in desalination plants is declining because of the stress corrosion cracking (SCC) problem, which is discussed in Chapter 4.

TABLE 2.1 Minimum mechanical properties of austenitic cast irons at room temperature.

ASTM A439 DESIGNATION	UNS No.	0.2% PROOF STRESS (MPa)	TENSILE STRENGTH (MPa)	ELONGATION (%)
D-2	F43000	207	400	8
D-2B	F43001	207	400	7
D-2C	F43002	193	400	20
D-2M	F43010	207	448	NS
D-3	F43003	207	379	6
D-3A	F43004	207	379	10
D-4	F43005	NS	414	NS
D-5	F43006	207	379	20
D-5B	F43007	207	379	6

NS = None specified

2.2 Copper Alloys

There is a wide range of copper alloys, but only three types are commonly used in desalination plants. These are discussed below and their nominal compositions are shown in Table A.2.

2.2.1 Brasses

Copper alloys containing zinc, commonly in concentrations from 5–40%, are known as brasses. Other elements may be added to modify the properties, such as strength, machinability, or corrosion resistance. The strength of brasses increases as the zinc content increases, although larger increases can be obtained by adding additional elements.

Brasses containing up to about 37% zinc are single-phase alpha alloys, while alloys containing more zinc are two-phase, with both alpha and beta phases. The quantity of beta phase increases as the zinc content (or alloying additions with a high zinc equivalence) increases. Some elements, such as silicon and aluminum, have a high zinc equivalence, so the beta phase will form at substantially lower zinc contents than with simple binary alloys. Aluminum brass, with 2% aluminum, has a zinc content of 22%, rather than 30%, to retain its all-alpha structure, which can then be protected from dezincification by arsenic additions (see Chapter 4).

Before the discovery of arsenic's ability to inhibit dezincification, it was found that tin reduced the rate of dezincification of the alpha phase. This is why alloys like naval brass contain 1% tin. The tin addition also improves the machinability of naval brass. Aluminum was found to increase the resistance of alpha brass to erosion corrosion, but at concentrations much above 2%, there was an increased susceptibility to localized pitting, particularly in polluted seawater.[2]

Arsenic is a required addition to many single-phase brasses to inhibit dezincification, as discussed in detail in Chapter 4. Most specifications require 0.02–0.06 wt% arsenic. The presence of manganese or iron impurities can "tie-up" part of the arsenic, so that it no longer acts as an inhibitor for dezincification.[3] Because dezincification resistant (DZR) brass often uses large quantities of scrap in the melting process, it was necessary to increase the arsenic range for this alloy to 0.02 to 0.15 wt%, with a typical concentration being 0.08 wt%.

The brasses traditionally used for hot stamping of fittings, etc., are duplex because all alpha-phase alloys tend to crack at forging temperatures. However, these duplex fittings are susceptible to dezincification in seawater. DZR brass was invented[3] to provide a brass that is two phase at hot stamping temperatures, but can be converted to an all-alpha structure by a simple, post-stamping heat treatment, typically 500 °C for 2 h. The presence of arsenic then makes the alloy resistant to dezincification. Campbell and Francis demonstrated that DZR brass is immune to dezincification in seawater,[3] and the alloy has been approved by Lloyds for hull fittings on marine boats, for example.

Lead is a common addition to brasses—usually 1–2%, which gives improved machinability. In addition, lead helps to make sand castings pressure tight. It is the lead addition to Muntz metal that aids its machinability.

"Bronze" is probably one of the most misused terms in metallurgy. Traditionally bronzes are alloys of copper and tin. However, some of the brasses have been termed "bronzes." Thus, 90/10 brass is sometimes called "commercial bronze." Another widely used term is "manganese bronze" for the family of high-tensile brasses. These alloys contain about 38% zinc, and 0.5–5% manganese with optional additions of iron, aluminum and lead. These are duplex brasses, which are heat treated to give moderate to high strength, with the strength increasing as the beta content of the alloy increases. Because these alloys are duplex, they are susceptible to dezincification and so service in aggressive waters, such as seawater, should be avoided, unless the alloy is galvanically protected. Alloys such as nickel aluminum bronze (NAB) will give equivalent or better strength combined with excellent corrosion resistance in seawater.

2.2.2 Copper-Nickel Alloys

Some common copper-nickel alloys are shown in Appendix A, Table A.2. Nickel and copper form a continuous solid solution so all the copper-nickel compositions are single phase, unless other elements are added. Manganese is usually present because it is used as a deoxidant in the melting process. Manganese also acts as a solid solution strengthener and increases the resistance to erosion corrosion. However, iron is five times more efficient at increasing the resistance to erosion corrosion[2] and this is why alloys such as 90/10 and 70/30 copper-nickel contain iron. With 30% nickel, iron is soluble up to several percent, but with 10% nickel, the solubility limit of iron at room temperature is about 1%. Hence, commercial 90/10 copper-nickel, with 1.5% or more iron, must be quenched from the annealing temperature (~ 900 °C) to keep the iron in solution. If this is not done, the corrosion resistance decreases significantly. Hence, it is important to check the quality of 90/10 copper-nickel with a simple magnetic permeability gauge to ensure the iron is in solution. This is discussed in more detail in Chapter 8.

The 66/30/2/2 Cu-Ni-Fe-Mn alloy was originally designed as a heat exchanger tube alloy with increased resistance to erosion corrosion in seawater containing high levels of silt or sand. It has very good resistance under these conditions,[2] but it has also been widely used with good success for brine heater tubes in multistage flash (MSF) desalination plants (see Chapter 6).

Copper-nickel alloys, particularly 90/10 and 70/30 copper-nickel, are widely used in desalination applications because of their excellent resistance to both fouling and corrosion. They are used as heat exchanger tubes, piping for seawater and firewater systems, and as cladding for larger steel items, such as water boxes.

The strength of the common copper-nickel alloys is only moderate (Table 2.2) and there are applications where a much higher strength is required, combined with corrosion resistance. One method of increasing the strength is to add aluminum (as in alloy C72400) and heat treat to precipitate gamma prime phase. After heat treatment, such alloys have high strength, as shown in Table 2.2. The principal use of these high strength alloys is for marine fasteners, where high strength and corrosion resistance is required.

Although this high strength alloy has good resistance to corrosion and erosion corrosion in seawater, it can suffer SCC under some circumstances, and in some fluids. This is discussed in more detail in Chapter 4.

TABLE 2.2 The minimum mechanical properties of some copper alloys at room temperature.

TYPE	COMMON NAME	UNS No.	0.2% PROOF STRESS (MPa)	TENSILE STRENGTH (MPa)	ELONGATION (%)
WROUGHT	Al-Brass	C68700	150	360	60
	Muntz Metal	C28000	160	380	40
	Naval Brass	C46400	170	400	35
	DZR Brass	C35330	120	280	40
	Mn-Bronze	C67500	195	460	15
	90/10 Cu-Ni	C70600	140	320	40
	70/30 Cu-Ni	C71700	170	400	40
	66/30/2/2 Cu-Ni-Fe-Mn	C71640	175	450	35
	Nibron	C72400	580*	800*	12*
	Al-Bronze D	C61400	230	540	40
	Al-Bronze E	C63000	290	710	20
	Nickel Aluminum Bronze	C63200	450	790	15
CAST	Admiralty Gunmetal	C92400	130	270	25
	Leaded Bronze	C93700	80	190	15
	LG2 Gunmetal	C83600	100	200	25
	LG4 Gunmetal	C92410	140	270	20
	M-Bronze	C92200	120	230	20
	Al-Bronze	C95200	170	500	40
	Nickel Aluminum Bronze	C95800	270	660	15

* Typical properties in the aged condition

2.2.3 Bronzes

Traditionally, the bronzes are alloys of copper and tin, with smaller additions of other elements for specific properties. The alloys known as "silicon bronze" and "aluminum bronze" have been named because they have many similarities with the tin bronzes in terms of metallurgical properties. Other alloys called bronzes, such as manganese bronze, commercial bronze, etc., are in fact, brasses and were discussed in Section 2.2.1.

The wrought tin bronzes usually contain small additions of phosphorus, used as a deoxidant in the melting process, and are sometimes called "phosphor bronzes." These alloys have very good corrosion resistance, particularly in seawater; the corrosion resistance increases as the tin content increases. The higher tin bronzes have very good resistance to polluted seawater (see Chapter 4). The wrought phosphor bronzes are often used for corrosion resistant springs.

When the tin bronzes are produced as castings, it is common to add some zinc and then the alloys are traditionally known as "gunmetals," because of their early use for cannon barrels. To ensure good pressure tightness of castings, used for valve bodies and pump casings, it is usual to add lead. This is why 10% phosphor bronze (C90700) may be used for pump impellers, but it is rarely used for pump cases, because the alloy contains no lead. "Admiralty gunmetal," also with 10% tin, offers the same corrosion resistance and it can be made pressure tight if lead is present at the upper level of the specification range (~ 2%).

LG2 gunmetal is a very old alloy that has been cast for over 200 years. However, it is difficult to cast in thick, complex sections and for such castings, LG4 gunmetal or M bronze is recommended.

If the lead content is increased further, to around 10%, the leaded tin bronzes make very good bearing alloys. For such applications, there is a long history of successful use in pumps and other rotating machinery.

Additions of tin in excess of ~ 10% result in the precipitation of delta phase. This makes the alloy even stronger, but also much more brittle and there are few commercial uses of such alloys.

The copper-aluminum alloys, or aluminum bronzes, combine high strength with good corrosion resistance and they are widely used in both wrought and cast forms, particularly NAB. For a detailed discussion of their metallurgy, properties, and corrosion resistance, the reader is referred to Harry

Meigh's excellent book on this class of copper alloy.[4] An extensive review of the corrosion resistance of aluminum bronzes was prepared by Campbell.[5]

Up to about 8% aluminum, the alloy is single phase. With higher aluminum contents, the beta phase is formed, which has good corrosion resistance and superior erosion corrosion resistance to the alpha phase. However, if these two-phase alloys are cooled too slowly, the gamma-2 phase can form and this phase is highly susceptible to corrosion. However, iron contents of 2% or more suppress the formation of gamma-2 phase and refine the grain structure.

When the aluminum content is 9–10%, additions of nickel and iron of 4–5% produce an alloy with high strength and corrosion resistance. The structure of the alloy is complex and was explained by Lorimer et al.[6] A simplified diagram of the structure is shown in Figure 2.1. As the alloy cools, the martensitic beta phase transforms to alpha and kappa phases. Figure 2.1 shows four kappa phases, all of which are enriched in iron and/or nickel. If the alloy cools too quickly, such as occurs during welding or with thin sections, not all the beta phase may transform. Residual beta phase (labeled "martensitic phase" in Figure 2.1) is anodic to the alpha phase and it will corrode rapidly in seawater.[2] However, a short heat treatment will ensure that all the beta phase transforms to alpha and kappa. For castings, heating to 675 ± 25 °C for 4–6 hours followed by air cooling is recommended. For wrought products a higher temperature of 725 ± 25 °C is preferred. For this reason, NAB should not be used in the as-welded condition in seawater and Royal Navy specifications forbid it (see Chapter 8).

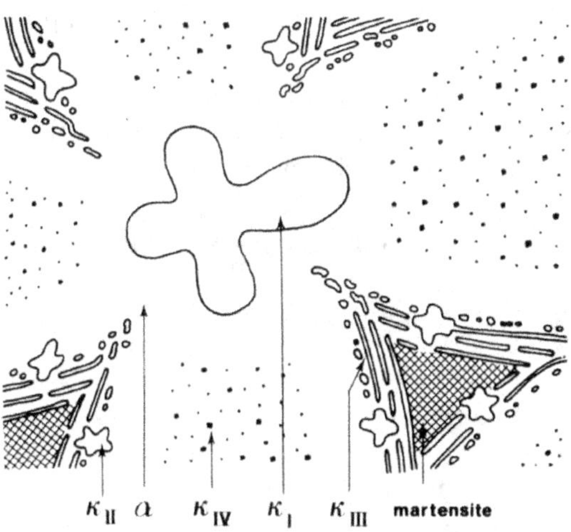

FIGURE 2.1 Schematic diagram of the structure of nickel aluminum bronze.[6]

The kappa III phase can cause corrosion problems if it becomes anodic to the alpha phase, because it is present in long thin stringers.[7] The origin of this problem and the best ways of avoiding it are discussed in Reference 2, and in Chapter 4.

Generally, the corrosion resistance of the aluminum bronzes increases as the aluminum, and other alloying additions increase. Hence, NAB in both wrought and cast forms is highly corrosion resistant. However, NAB is very susceptible to erosion corrosion in polluted seawater (Chapter 4) and in such applications, high-tin bronzes are preferred. In addition to clean seawater, NAB has excellent resistance to corrosion in a wide range of aggressive chemicals,[5] including hydrofluoric acid and fluorine.

2.3 Stainless Steels

A stainless steel is one containing more than 11% chromium, and many contain much more. It is the chromium that confers corrosion resistance and the highest alloyed stainless steels are resistant to corrosion in a wide range of chemicals. Stainless steels contain carbon, but many modern grades contain 0.03% or less, which greatly reduces the risk of carbide formation during heating. Low carbon gives good weldability without loss of corrosion resistance, because carbides tie up chromium from the matrix. The compositions of some commonly used stainless steels are shown in Tables A.3 (wrought) and A.4 (cast).

Stainless steels often contain additions of other elements, which can change the metallurgical structure, such as nickel (as described in the following sections), or they can change the corrosion resistance, such as molybdenum. They may also contain other elements to ease manufacture or increase resistance to a specific chemical.

The corrosion resistance of different grades is often compared using the pitting resistance equivalent number, or PREN, where.

$$\text{PREN} = \%\text{Cr} + 3.3(\%\text{Mo} + 0.5\times\%\text{W}) + 16\times\%\text{N}.$$

This is an empirical formula that assesses the resistance of a stainless steel (or Ni-Cr-Mo alloys) to localized corrosion, such as pitting or crevice corrosion, in a near-neutral chloride environment, like seawater. The higher the PREN the greater is the resistance to localized attack. The formula is widely used in industry and has been adopted by ISO 15156/NACE MR0175, the standard for materials use in sour oil and gas service.[8] It has been widely demonstrated that a PREN > 40 is required to give resistance to crevice corrosion in warm seawater, and this requirement for seawater service has been adopted into some standards, such as NORSOK, the Norwegian oil and gas standard.[9]

Kovachs et al.[10] tested a range of stainless steels and found that to resist crevice corrosion in warm seawater an alloy required a critical crevice temperature (CCT) of ≥ 35 °C in the ASTM G48 test (Figure 2.2). When they plotted the CCT against the PREN for a range of alloys they got separate lines for ferritic, duplex, and austenitic stainless steels (Figure 2.3). The results showed that a ferritic alloy needed a PREN > 35, a duplex alloy > 40, and an austenitic alloy > 43 to resist crevice corrosion in warm seawater.

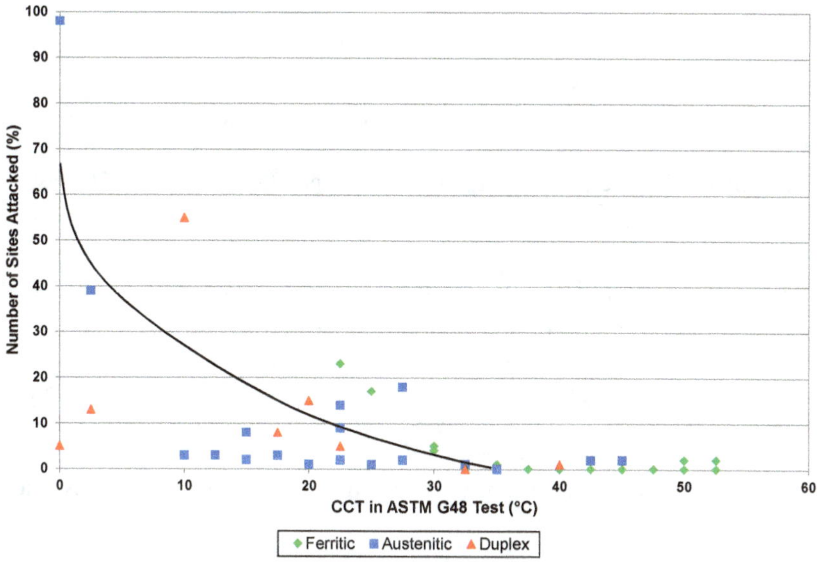

FIGURE 2.2 Percentage of crevice sites attacked in warm seawater versus the CCT in the ASTM G48 test for some stainless steels.[10]

More recently there are newer methods available for manufacturing stainless steels that are particularly suited to complex shapes. The first is hot isostatic pressing (HIP) in which fine powder is placed in a mild steel mold, which is then sealed and evacuated (ASTM A988[11]). The whole item is then heated at high temperature and pressure to produce a 100% dense solid that is isotropic. The steel outer can then be pickled or machined off. It is known that

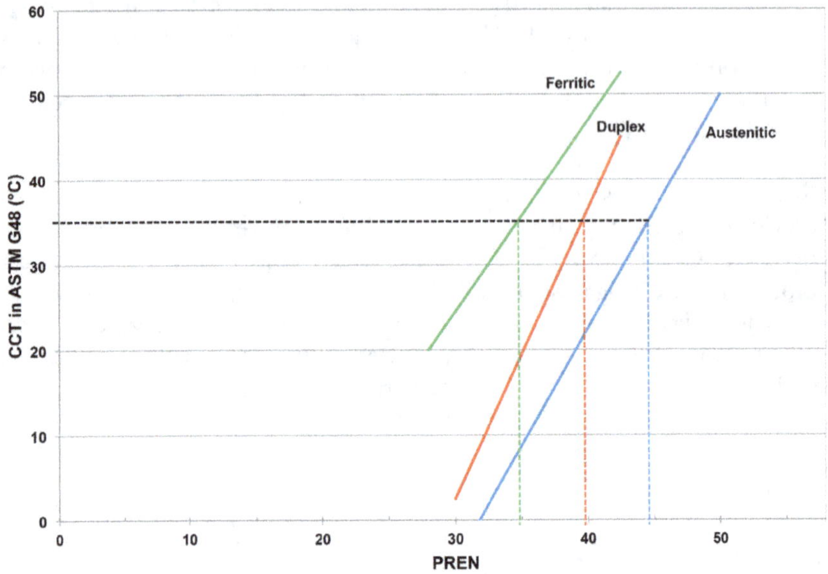

FIGURE 2.3 The PREN of some stainless steels versus CCT in the ASTM G48 test.[10]

HIP duplex products do not always have the same corrosion resistance as conventional material. The second method is 3-D printing, which currently is limited to small and medium size parts. At present there is no published data on corrosion of 3-D printed stainless steels in desalination environments. It is recommended that alloys manufactured by either of these methods be subjected to corrosion tests prior to use, because it is not safe to assume equivalent corrosion resistance to conventional wrought and cast products.

2.3.1 Ferritic and Martensitic Stainless Steels

Ferritic stainless steels retain the FCC structure of ferrite, which makes them strong, but difficult to manufacture and weld, particularly in thick sections. Alloys with around 12 or 13% chromium should not be used in seawater, and are not very resistant to aggressive marine atmospheres. There are ferritic alloys with 26% or more chromium that are very resistant to crevice corrosion in seawater, such as Sea-Cure (Plymouth Tube Company, East Troy, Wisconsin) and Al 29-4C (Allegheny Ludlum, Pittsburgh, Pennsylvania). Some typical compositions are shown in Appendix A, Table A.3. It can be seen that both these super ferritic alloys have a PREN > 35, as discussed above, for resistance to crevice corrosion.

Ferritic stainless steels are difficult to manufacture in thick sections, particularly the higher alloyed grades. Hence, Sea-Cure and AL 29-4C are usually used only as heat exchanger tube, where they are rolled into galvanically compatible tube plates, because of the difficulty in welding these alloys (see Chapter 3).

Martensitic stainless steels contain 11.5–18% chromium and 0.1–1% carbon so that they are austenitic at elevated temperatures, and this can be transformed to martensite by suitable cooling to room temperature. Martensite is both hard and strong, but also has low ductility. Hence, martensitic stainless steels suitable for practical engineering must be tempered (heat treated) to reduce the martensite content and recover some ductility. Some martensitic stainless steels may also contain additions of nickel and/or molybdenum to modify the properties.

Martensitic stainless steels have chromium contents in the range 11.5–18% and thus do not have very good resistance to chlorides. They suffer rapid localized corrosion in seawater and are also susceptible to corrosion in marine atmospheres. They are not generally used in desalination plants.

2.3.2 Austenitic Stainless Steels

Austenitic stainless steels have an FCC structure, which is stabilized by adding an austenite-forming element, such as nickel and/or manganese. Nickel is most commonly used because high concentrations of manganese reduce the ductility, and the corrosion resistance in seawater. Nitrogen is also an austenite former and it is added, particularly to the high-alloy stainless steels, where it not only stabilizes the austenite structure, but it also increases strength, improves weldability and corrosion resistance, as well as improving the hot forming characteristics. The typical compositions of some common wrought austenitic alloys are shown in Appendix A, Table A.3.

Where austenitic alloys, particularly the 300 series, are made to the low nickel end of the composition range, it is possible to have some residual delta ferrite in the structure. This can range from 5–20%, but is usually at the lower end of the range. At this concentration, it does not have a significant effect on the properties or the corrosion resistance. It has been reported that ferrite contents at the higher end of the range can increase the resistance to chloride stress corrosion cracking, but this has not been quantified.

The most well-known austenitic grades are 304 and 316, in particular the low carbon L grades, which makes them readily weldable. They do not resist crevice corrosion in seawater, but they are widely used in low chloride waters at moderate temperatures for a wide range of applications (see Chapter 4). They have also been used in aerated seawater with cathodic protection, for items such as tube sheets, and also in low oxygen areas of thermal desalination plants (see Chapters 6 and 7).

While the L grades are preferred for wrought products, because of ease of welding without corrosion issues, castings may either use the standard grades–304 and 316 (0.08C max.)–or the L grades–304L and 316L (0.03C max.). Although the minimum strength in ASTM A351 (Table 2.4) is the same for both grades, it is usual to obtain slightly higher strength with the higher carbon grade. Smaller pumps and valves tend to be the L grade, particularly if they must be welded into piping. Larger pump cases usually have cast-in flanges for piping connections, but weld repair may be needed if there are minor defects, such as porosity, in the castings. Where there is concern about welding, specifying CF3M is usual, but it has been known to specify 316 castings (CF8M), with a maximum carbon of 0.05%. This will allow weld repair without undue problems.

The corrosion resistance of the austenitic alloys is increased by adding more chromium and molybdenum, as shown by alloy 904L. This alloy can handle higher chloride waters compared with 316L, but it is not seawater resistant (see Chapter 4). Only the 6% Mo austenitic alloys, with a PREN >43, have good resistance to crevice corrosion in warm seawater. These alloys are available in a wide range of product forms, and have been used for seawater piping, heat exchangers, and for cladding carbon steel.

The typical mechanical properties are shown in Table 2.3 and it can be seen that the strength increases with alloy content, while good ductility is retained by all grades.

Table A.4 shows the typical composition of some cast austenitic stainless steels that have been used in desalination plants. These are the cast versions of 316L and the 6% Mo alloys in Table A.3. The compositions are very similar to their wrought counterparts. There can be problems casting the 6% Mo alloys, particularly in thicker sections and these are discussed by Hebdon and Francis, who also suggest the use of cast superduplex alloys in difficult cases.[12]

The typical mechanical properties of the cast austenitic stainless steels are shown in Table 2.4, as taken from ASTM A351.[13] The strength of cast 316 is similar to the wrought product, but the cast 6% Mo alloys are not as strong as their wrought counterparts.

TABLE 2.3 The minimum mechanical properties of some wrought stainless steels.

TYPE	UNS No.	GENERIC NAME	0.2% PROOF STRESS (MPa)	TENSILE STRENGTH (MPa)	ELONGATION (%)
Austenitic	S31603 N08904	316L 904L	170 220	485 490	40 35
Super Austenitic	S31254 N08367 N08926	6%Mo 6%Mo 6%Mo	310 310 310	655 655 655	35 35 35
Ferritic	S41003	410	220+	414+	40+
Super Ferritic	S44735 S44660	Al 29-4C Sea-Cure	550 380	655 550	20 20
Lean Duplex	S32304 S32101 S82011 S32202	2304 2101 2102 2202	400 450 450 450	630 650 650 650	25 25 25 25
Duplex	S32205	2205	450	655	25
Superduplex	S32760 S32750	Z100 2507	550 550	750 750	25 25

+ In annealed condition (may be heat treated up to ~1,000 MPa proof stress and 17% elongation).

TABLE 2.4 The minimum mechanical properties of some cast stainless steels.

TYPE	ASTM+	GENERIC NAME	0.2% PROOF STRESS (MPa)	TENSILE STRENGTH (MPa)	ELONGATION (%)
Austenitic	CF8M CF3M CK3MCuN CN3Mn	316 316L 6%Mo 6%Mo	205 205 260 260	485 485 550 550	30 30 35 35
Standard Duplex	Grade 4A	2205	415	620	25
25% Cr Duplex	Grade 1A Grade 1B	CD4MCu CD4MCuN	485 485	690 690	16 16
Superduplex	Grade 5A Grade 6A	2507 Z100	515 450	690 690	18 25

+ ASTM properties taken from A351 for austenitic and A995 for duplex.

2.3.3 Duplex Stainless Steels

Modern duplex stainless steels are mostly produced with a roughly 50/50 austenite/ferrite phase balance. Any heat treatment or other process that takes the alloy too far from the nominal balance, or introduces third phases, can result in reduced mechanical properties, corrosion resistance, or both.

The duplex stainless steels can be split up into three main groups: lean duplex, standard duplex, and superduplex. The compositions of some common wrought alloys are shown in Appendix A, Table A.3, while Table A.4 shows the composition of some commonly used cast alloys. A more robust description of the full range of duplex stainless steels, including their compositions, properties, and corrosion resistance in a wide range of fluids can be found in Reference 14.

Steel mills have developed a range of lean duplex alloys, with low nickel and molybdenum to provide a higher strength alternative to 304L and 316L. The nominal compositions are shown in Table A.3. The earliest lean duplex was 2304, which provides a corrosion resistance close to that of 316L and better than 304L. More recently, alloys with less nickel and more nitrogen have been developed to provide similar corrosion resistance and high strength at lower cost. The minimum mechanical properties are shown in Table 2.3, where the higher strength compared with 316L is clear. Lean duplex alloys have been used with cathodic protection for tube sheets in thermal desalination plants, and they have been used in low oxygen environments instead of 316L (see Chapters 6 and 7). The three alloys–2101, 2102 and 2202–are all proprietary with similar properties. They are lower cost than 316L, but the factor holding up their wider use, at the time of writing, is the limited availability in product forms such as pipe, fittings, flanges, and bar. They are all available as plate and sheet.

Alloy UNS S31803 is the original specification for 2205 duplex, and UNS S32205 was introduced to guarantee a consistently high level of corrosion resistance with higher minimum contents for chromium, molybdenum, and nitrogen. UNS S32205 is now the preferred composition and most manufacturers dual certify their 2205 to both specifications. Alloy 2205 is widely used in many industries, particularly where the SCC resistance of the 300 series austenitic alloys is inadequate. It offers a higher strength and lower cost than the austenitic alloy 904L, with a similar corrosion resistance in many fluids.

The superduplex stainless steels contain 25% chromium and offer a similar corrosion resistance or better when compared with the 6% Mo austenitic alloys, but at lower cost, because of their lower nickel and molybdenum contents. These alloys have found a wide variety of uses in aggressive process fluids, as well as seawater and sour oil and gas fluids.

The cast alloys are a little different, in that there is not such a wide variety to choose from. Only six alloys are listed in ASTM A995 (for pressure-containing castings).[15] The compositions of some common cast alloys are shown in Table A.4.

ASTM A995 grade 4A is the cast version of 2205 and this alloy has been widely used for pumps, valves, and other cast components in environments less aggressive than seawater. CD4MCu is an older 25% Cr cast alloy that was originally developed for sulfuric acid service, but the modern version, CD4MCuN, contains nitrogen. This gives the alloy an improved resistance to chlorides, as well as better weldability, and a more equal phase balance. CD4MCu is no longer listed in ASTM A995 for duplex castings.

Of the superduplex alloys, Z100 (grade 6A) started life as a casting alloy and has been widely used for many cast components, including pumps and valves. Care is required in casting when producing thick section components, but castings up to 5 tonnes finished weight have been produced successfully. The cast version of 2507 (grade 5A) has a slightly higher strength than Z100 because of its higher ferrite content. This can lead to cracking problems when trying to produce thicker section components.[12]

The mechanical properties of the duplex stainless steels are shown in Tables 2.3 (wrought) and 2.4 (cast). It can be seen that the duplex stainless steels are stronger than austenitic alloys of similar PREN. For almost every austenitic alloy, there is a duplex stainless steel with similar corrosion resistance, higher strength and, often, lower cost. This higher strength means that duplex alloys have higher design stresses than austenitic alloys for vessels and piping, as discussed in References 14 and 16, for example. This means that there can be savings in both material and fabrication costs by designing to the strength of duplex stainless steels.

2.4 Nickel Alloys

Pure nickel is rarely used in engineering, apart from in the chlor-alkali industry because it has very good resistance to hot, concentrated caustic soda. Nickel is usually alloyed with other elements and the choice of these depends on the required properties. The most common combinations for engineering alloys are nickel-copper, nickel-molybdenum, and nickel-chromium-molybdenum. The nickel-molybdenum alloys are not used in the desalination industry, but the other two classes of nickel alloys have some applications and are described below.

2.4.1 Nickel-Copper Alloys

The nickel-copper alloys have been around for about 100 years and the most common one is alloy 400, whose composition is shown in Appendix A, Table A.5. Alloy 400 has good resistance to chloride-containing solutions in the pH range 5–14, but it is not suitable for use in oxidizing, acidic solutions. It has been widely used in seawater systems for over 50 years, often with excellent results.

Where higher strength is required, a precipitation hardened version of the alloy, K-500, is available. This can be age-hardened to a range of strengths by the formation of gamma prime phase. Table 2.5 shows the typical mechanical properties of both alloys. Alloy K-500 can be heat treated to higher strengths, with a corresponding loss in ductility. Its primary use has been for high strength, corrosion resistant fasteners, pump shafts, and valve stems. Alloy K-500 cannot be welded without losing its high strength (see Chapter 3).

The 6% Mo austenitic and superduplex stainless steels often have the same corrosion resistance in specific solutions as these nickel-copper alloys and for lower cost.

2.4.2 Nickel-Chromium-Molybdenum Alloys

These alloys tend to be used in aggressive environments when stainless steels do not have adequate resistance to corrosion. Table A.5 in Appendix A shows a range of compositions for some of the commonly used alloys. As with stainless steels, their corrosion resistance increases as the PREN increases. Nitrogen is not added to these alloys, because it cannot fit easily into the FCC nickel matrix, unlike the larger iron matrix in high-alloy austenitic stainless steels. However, many of these alloys contain large quantities of molybdenum or tungsten and so the PREN ranges widely from 31 for alloy 825, to 79 for alloy 686.

Alloy 825 has a relatively low PREN, and it is not resistant to crevice corrosion in seawater, but it has excellent corrosion resistance under reducing conditions. Alloy 625 contains more molybdenum, which means that its resistance to chlorides is very good and it can be used even when the pH de-

TABLE 2.5 The minimum mechanical properties of some common nickel alloys at room temperature (taken from ASTM standards for hot rolled plate).

NAME	UNS No.	0.2% PROOF STRESS (MPa)	TENSILE STRENGTH (MPa)	ELONGATION (%)
Alloy 400 Alloy K-500	N04400 N05500	195 690*	485 965*	35 20*
Alloy 825	N08825	241	586	30
Alloy 625	N06625	379	758	30
Alloy C-276	N10276	283	690	40
Alloy C-22	N06022	310	690	45
Alloy 59	N06059	310	690	45
Alloy 2000	N06200	310	690	45
Alloy 686	N06686	283	690	45
Alloy 718	N07718	1034*	1275*	12*
Alloy 925	N09925	758*	965*	25*
Alloy 725	N07725	827*	1137*	20*

*Age hardened. The properties in this condition can vary and typical values are shown.

creases. Alloy C-276 has been a workhorse alloy for the chemical and process industries for many years because of its resistance to a wide range of corrosive fluids, including seawater. Alloys C-22, 2000, 59, and 686 are all more highly alloyed versions of C-276, with even greater corrosion resistance in aggressive fluids. Although alloy C-22 has a slightly lower PREN than C-276, its higher chromium content can give it better performance than C-276 in some environments. These higher nickel alloys have been widely used for plate heat exchangers, particularly when seawater is the coolant. In addition, they have been used as weld overlay onto carbon or stainless steels in seawater service in areas at high risk of accelerated corrosion.

These alloys are higher cost than stainless steels and are usually selected for aggressive corrosive services. However, under some conditions, the higher alloyed stainless steels offer similar performance to nickel alloys at lower cost.

All of these nickel alloys have been cast, more or less to the same composition as the wrought version.

There are also precipitation hardened nickel-chromium-molybdenum alloys, which can be aged to high strength. The composition of some common alloys is shown at the bottom of Table A.5 in Appendix A (alloys 718, 925 and 725).

All these alloys contain additions of aluminum and titanium, and increase their strength by the precipitation of gamma prime phase as the alloys are progressively aged. Alloy 925 is the age hardening version of alloy 825, while alloy 725 is the age hardening version of alloy 625.

Alloys 718 and 925 have relatively low PREN, and, hence, limited resistance to chlorides, although alloy 925 has been used for marine fasteners in aggressive atmospheres. Alloy 725 has a greater resistance to chlorides because of its higher molybdenum content, similar to that of alloy 625, and it has found uses in seawater where high strength is required.

Some typical mechanical properties in the aged condition are shown in Table 2.5. The strength can be increased further with additional aging, at the cost of a reduction in ductility. Alloy 725 has been aged to give a 0.2% proof stress in excess of 1,000 MPa, while still retaining some ductility. It is important to balance corrosion resistance and strength when using these alloys.

2.5 Titanium and Its Alloys

Titanium and its alloys are a class of materials with outstanding resistance to corrosion in neutral and oxidizing solutions over a wide range of pH values. The use of these alloys under reducing conditions, however, requires caution. Although they are more costly than many other materials, their outstanding performance under arduous conditions has shown titanium and its alloys to be cost effective solutions in some desalination applications.

Table A.6 in Appendix A shows the compositions of some titanium alloys commonly used in the desalination industry. Titanium grades 1 and 2 are probably the most widely used. Grade 1, with lower oxygen, has better cold formability and is preferred for such products as titanium clad steel plate, and plates for plate and frame heat exchangers. Besides increasing ductility at room temperature, low oxygen (0.13% max) also improves toughness and confers higher fatigue crack growth resistance.

Grade 2 titanium has a slightly higher oxygen content than grade 1, and this grade is widely used for seamless and seam welded heat exchanger tubing. The addition of ruthenium or palladium further increases the corrosion resistance of titanium, as do small additions of molybdenum and nickel. These alloys are listed in Table A.6, but they are not normally required in desalination plants, even under the most demanding applications.

All of the all-alpha phase alloys have moderate strength, as shown in Table 2.6. The alpha/beta alloys have higher strength, as shown in Table 2.6 and the Ti/6Al/4V alloy has also been made as castings. The wrought product has been used for high strength fasteners and valve stems, while the cast version has been used for the bodies of pumps and valves. However, the high cost of the cast product means that high-alloy stainless steels are usually more cost effective.

TABLE 2.6 The minimum mechanical properties of some common titanium alloys at room temperature (taken from ASTM standards for hot rolled plate).

TYPE	GRADE	UNS No.	0.2% PROOF STRESS (MPa)	TENSILE STRENGTH (MPa)	ELONGATION (%)
Alpha	1	R50250	170	240	24
	2	R50400	275	345	20
	3	R50550	380	450	18
Alpha/Beta	5	R56400	828	895	10

The exceptional corrosion resistance of titanium alloys in aggressive corrosive environments originates from the formation of a thin (typically 50–300 Å thick) highly chemically stable, adherent, ceramic-like, protective surface oxide film. This tenacious oxide film is primarily TiO_2, and forms spontaneously and instantaneously when fresh metal surfaces are exposed to traces of oxygen (air) or water (moisture) alone. As such, a mechanically damaged (e.g., scratched) oxide film can instantly repair/heal itself in air or submerged in aerated or fully deaerated solutions containing water. The stability/formation of this TiO_2 film is not inhibited/affected by the presence of chlorides, and can occur over a wide range of pH depending on temperature (i.e., pH 2–12 at room temperature).[5]

Although titanium is very resistant to localized corrosion in hot, chloride solutions at low pH, it does have limits, but the conditions in desalination plants do not generally extend to the limits of the titanium alloys grades 2 and 5 in Table A.6.

One ion that has a marked deleterious effect on the corrosion of titanium and its alloys is fluoride, and this is discussed in Chapter 4.

2.6 Polymers

These are used extensively in desalination applications for both chemical storage vessels and piping to handle both corrosive chemicals and seawater. Their combination of chemical resistance to a wide range of chemicals, light weight, and relatively low cost (compared to Ni-Cr-Mo alloys and high-alloy stainless steels), make them a very attractive option.

2.6.1 Fiberglass Reinforced Plastic (FRP)

Fiberglass reinforced plastics (FRPs) are made from thermoset polymer resins reinforced with glass fibers. The glass fibers provide the strength while the thermoset resin provides a corrosion resistant binder. While the different resins used for FRP have different mechanical properties, the strength of the laminate is governed by the glass content, orientation of the fibers, and method of construction.

The structural layer of the FRP laminate is composed mainly of glass to provide a high strength composite material. The surface of the laminate exposed to the process is polymer rich and is known as the "corrosion liner" or "corrosion barrier." Only sufficient reinforcement from either glass or polyester is required to strengthen the corrosion liner (Figure 2.4).

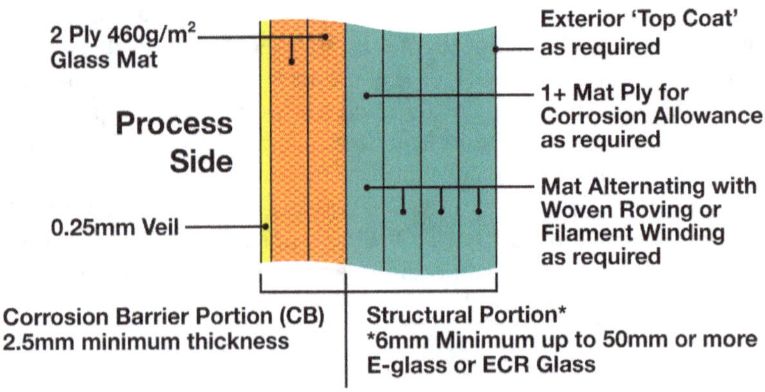

FIGURE 2.4 Schematic diagram showing make up of FRP.

Epoxy resins are the least expensive of the polymers that are used to make FRP composites. Their low cost makes them very attractive for a number of applications. They are widely used to handle seawater, particularly as large diameter feed pipes.

Other resins are sometimes used where the liquid is particularly corrosive, the temperatures are higher, or better mechanical properties are required:

- Epoxy vinyl ester resins (higher chemical resistance)
- Epoxy novolac-based vinyl ester resins (higher temperature limit)
- Brominated epoxy vinyl ester resins (good fire resistance and chemical resistance)
- Brominated epoxy novolac vinyl ester resin (better than brominated epoxy vinyl ester resin)
- Elastomer-modified bisphenol-A epoxy vinyl ester resin

The latter resin is typically used when bonding other FRP resins to steel or other substrates, such as concrete or wood. Their semi-elastomeric behavior, when used as an interlayer, accommodates the different thermal expansion coefficients of the FRP and the substrate.

Equipment and piping made from FRP have proven to be extremely well suited to many corrosive environments. Vinyl ester resins are immune to attack by many chemicals, brines, and weak acidic solutions. FRP piping can be used in caustic applications, but care must be taken to prevent caustic attack of glass reinforcement. For this reason, FRP pipe for caustic applications is specified with a "synthetic" veil, which is typically polyester. Some operators have required carbon veils for caustic service, but this is not usually necessary because the polyester veil is usually sufficiently resistant. Vinyl ester resins are capable of handling sodium hydroxide at a range of concentrations up to 25% at up to 65 °C if the article (pipe or vessel) is post-cured to ensure full polymerization and a synthetic veil is used.

Chemical resistance for FRP is best assessed on a case-by-case basis. Resin suppliers publish chemical resistance guides for their products and these can be extremely useful for selecting FRP resins for commonly used chemical environments.[17] The makeup of the corrosion liner will determine the chemical resistance and therefore the long-term performance of the laminate. The variables that dictate the makeup of the corrosion liner are

- resin type
- composition of the reinforcement (glass type)
- composition of the outermost sheet in the corrosion liner (veil)
- thickness of the corrosion liner
- type of curing system used to polymerize the resin

The thickness and makeup of the corrosion liner is best assessed by an experienced person familiar with the chemical system and the types of resin that can be used in the corrosion liner. In many cases, testing of representative laminates with candidate corrosion liners can provide useful materials selection data.

2.6.2 Polyvinyl Chloride (PVC)

PVC and chlorinated PVC (CPVC) are variants of the same polymer backbone. PVC is made by the substitution of one of the hydrogen atoms on the ethylene monomer repeat unit with a chlorine atom. The resulting polymer is more chemically resistant, stiffer, and stronger than polyethylene. The mechanical properties of PVC can be changed significantly by the addition of plasticizers. Plasticized PVC can be made to be very flexible and has a wide range of applications in the packaging, clothing, and other industries. The grades of PVC used in the chemical and desalination industry are "unplasticized" or "uPVC." This material is widely used for handling chemicals including hydrochloric acid across its concentration range (up to 37%). PVC can be chlorinated by taking the PVC resin and exposing it to chlorine under heat and pressure to produce chlorinated PVC. CPVC has improved mechanical strength over uPVC and is often used where the temperature for uPVC is too high. Typically, the limiting temperature for PVC is 60 °C, whereas the limiting temperature for CPVC is 93 °C. Figure 2.5 is a useful comparison of the limiting temperatures at different pressures for thermoplastic piping materials.[18] Note that in terms only of mechanical properties, CPVC outperforms most thermoplastics and even outperforms polyvinylidene fluoride (PVDF), a fluoropolymer, over most of the temperature range.

Another issue is the brittleness of the thermoplastics. If a worker drops a wrench or other heavy object on a PVC pipe it can shatter and empty potentially corrosive fluids over the site. For this reason, it is common to apply a thin FRP "wrap" around the thermoplastic pipe as armoring against mechanical damage. It is advisable to carry out a drop test on thermoplastics to ensure they are not excessively brittle.

2.6.3 UV Degradation of Polymers

Most polymers will suffer degradation if exposed to strong sunlight with a high ultraviolet content, as occurs in tropical climates and the Middle East. The effects of this can be minimized by taking suitable precautions.

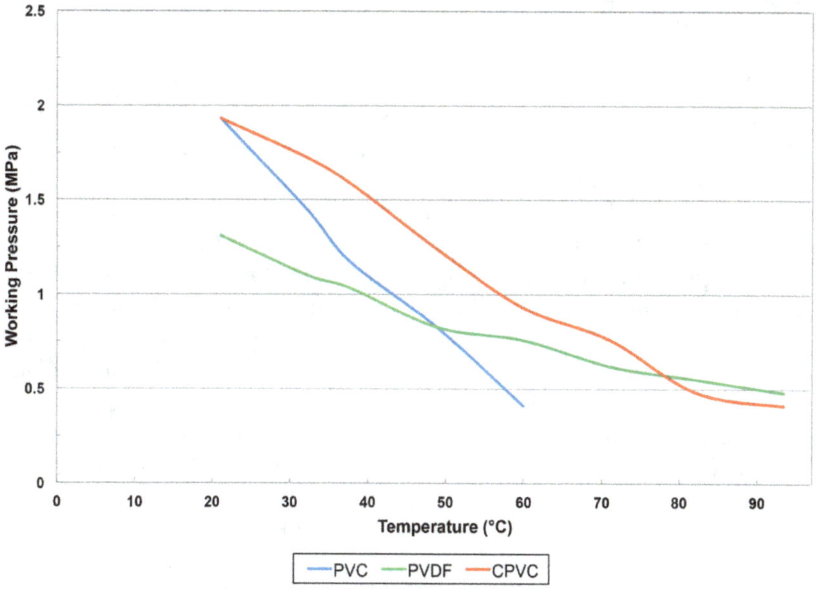

FIGURE 2.5 Maximum working pressures for some polymer piping materials.

It is important to specify polymer grades with UV protection built in. However, in very hot climates this is not sufficient to give extended life and a topcoat of white latex paint acts as a sunblock and also tends to reflect solar radiation. Many end-users in hot/tropical climates tend to replace their uPVC and CPVC pipe on a timed life basis not only because of chemical attack, but also because of UV issues.

Many chemical suppliers supply their chemicals in totes, which are high-density polyethylene (HDPE) in a steel cage, and these will also need protection from UV if the chemical is dispensed directly from such containers.

Where polymers have failed by UV degradation, it is often because a grade without UV protection has been used outdoors, or a supplier has provided a grade without the UV protection.

REFERENCES

1. ASTM A439 (latest revision), "Standard Specifications for Austenitic Ductile Iron Castings" (West Conshohocken, PA: ASTM International).
2. Francis, R., *The Corrosion of Copper and Its Alloys: A Practical Guide for Engineers* (Houston, TX: NACE International, 2010).
3. Campbell, H.S., and R. Francis, "Dezincification Resistant Brass, CZ132, in Seawater," *The Metallurgist* (April 1983).
4. Meigh, H.J., *Cast and Wrought Aluminium Bronzes, Properties, Processes and Structures* (London, United Kingdom: Institute of Materials, 2000).
5. Campbell, H.S., "Aluminium Bronze Alloys Corrosion Resistance Guide," Publication No. 80 (Hertfordshire, United Kingdom: Copper Development Association, 1981).
6. Lorimer, G.W., F. Hasan, J. Iqbal, and N. Ridley, "Observation of Microstructure and Corrosion Behaviour of Some Aluminium Bronzes," *British Corrosion Journal*, 21 (1986): pp. 244–248.
7. Rowlands, J., "Studies of the Preferential Phase Corrosion of Cast Nickel Aluminium Bronze in Seawater," 8th International Congress on Marine Corrosion and Fouling (1981): pp. 1346–1351.
8. NACE MR0175/ISO 15156 (latest revision), "Petroleum and Natural Gas Industries, Materials for Use in H_2S-Containing Environments in Oil and Gas Production, Part 3. Corrosion Resistant Alloys" (Houston, TX: NACE International).
9. NORSOK M-001, "Materials Selection," Edition 5 (Oslo, Norway: Standards Norway, 2014).
10. Kovach, C.W., and J.D. Redmond, "Correlations Between the Critical Crevice Temperature, PRE-Number and Long Term Crevice Corrosion Data for Stainless Steels," CORROSION/93, paper no. 276 (Houston, TX: NACE International, 1993).
11. ASTM A998 (latest revision), "Standard Specification for Hot Isostatically Pressed Stainless Steel Flanges, Fittings, Valves, and Parts for High Temperature Service" (West Conshohocken, PA: ASTM International).
12. Francis, R., and S. Hebdon, "The Selection of Stainless Steels for Seawater Pumps," CORROSION 2015, paper no. 5446 (Houston, TX: NACE International, 2015).
13. ASTM A351 (latest revision), "Standard Specification for Castings, Austenitic, for Pressure-Containing Parts" (West Conshohocken, PA: ASTM International).
14. Francis, R., *The Corrosion of Duplex Stainless Steels: A Practical Guide for Engineers* (Houston, TX: NACE International, 2018).
15. ASTM A995 (latest revision), "Standard Specification for Castings, Austenitic-Ferritic (Duplex) Stainless Steel, for Pressure-Containing Parts" (West Conshohocken, PA: ASTM International).

16. Byrne, G., R. Francis, G. Warburton, R.J. Bullock, and C. Kuzler, "The Selection, Design, Fabrication and Performance of Zeron 100® in SWRO Applications," IDA Conference (Topsfield, MA: International Desalination Association, 2003).
17. Ashland Chemical, "Resin Selection Guide for Corrosion Resistance," https://www.ashland.com/file_source/Ashland/Industries/Corrosion%20Resistant/Related%20Links/Chemical%20Resistance%20Guide.pdf (December 27, 2018).
18. Clarke, S., and R. Francis, *The Selection of Materials for Mineral Processing Operations* (Houston, TX: NACE International, 2016).

3 Fabrication

In this chapter, the welding characteristics of the materials commonly used in desalination plants are discussed, along with the effects of welding on corrosion resistance. Where the reader requires further information, they are referred to the references in the relevant section.

Welding is one of the most common methods of joining, and the composition and structure of the weld metal may be quite different from that of the parent metal.[1-2] It is important, therefore, that the weld metal is not galvanically more susceptible to corrosion than the parent metal. Because the weld metal area is relatively small compared with the adjacent parent metal, any attack at the weld is greatly accelerated (see Chapter 4 on Galvanic Corrosion).

3.1 Carbon Steel

With carbon steel, problems of preferential corrosion of the weld metal usually result from the choice of welding consumables. High-silicon, low-nickel consumables give welds that are rapidly corroded in seawater, while the reverse (i.e., low-silicon, high-nickel) can produce welds that are acceptable. Figure 3.1 shows severe corrosion of a carbon steel weld in low oxygen (~100 ppb) seawater, which was 50 mV more negative in potential compared with the parent pipe. A common filler used to avoid preferential weld corrosion contains 0.6–1.0% nickel and 0.4% copper.[1]

Another problem that sometimes occurs with carbon steel is preferential corrosion of the heat-affected zone (HAZ). If the welding is carried out with a heat input that is too low, then the HAZ can be hard and develop preferential corrosion in some aqueous media at room temperature. The solution is to weld with a higher heat input to produce a softer HAZ. This can be a problem with some alloyed steels, because welding produces a more coarse grain size than the grain size in the parent metal.[2]

For more information, S. Bond (Ref. 1) presented a paper reviewing the corrosion of carbon and low-alloy steel welds, the causes of problems, and the methods of avoidance.

FIGURE 3.1 Preferential corrosion of a carbon steel weld in piping carrying partially deaerated seawater (~ 100 ppb oxygen). [Photo courtesy of Intertek]

3.2 Austenitic Cast Iron

Austenitic cast iron is difficult to weld and when it is welded, it is usually done by an experienced fabrication shop. It is unlikely that staff doing site weld repairs would have the necessary experience to do it properly.

Even with an experienced fabricator in a welding shop, it is not easy and the results can be variable. Generally the best results are obtained with manual metal arc (MMA) welding and Ni/Fe electrodes. A preheat around 100 °C is required with an interpass temperature of 100–150 °C. After welding, it is important to cool the weld slowly, either by insulation wrapping or a furnace cool. In thicker sections, the welding technique is important and buttering combined with careful balancing of passes is recommended.

Because of the risk of SCC in warm seawater (see Chapter 4) it is recommended that the component be stress relieved after welding. This is usually done at around 450–550 °C, followed by slow cooling, as described in Chapter 2.

3.3 Copper Alloys

When welding copper alloys, sometimes the joints are autogenous, and sometimes filler is used, which is, often of a matching composition to the parent alloy. For some of the copper alloys, the selection of a filler metal is very important. Copper and its alloys can be welded by all of the common arc welding techniques. It is important that the weld metal is not anodic to the parent metal, because the weld bead has a much smaller area than the parent metal, so it is important that its potential is either the same as the parent metal or more positive. If this is not achieved, rapid, localized attack of the weld metal can occur.[2]

The copper-nickel alloys are often used in seawater and welding is preferred for high strength joints. The iron content of 90/10 copper-nickel exceeds the solubility limit at room temperature (see Chapter 2) so iron may come out of solution if a matching composition filler is used and the weld cools too slowly. This can reduce the corrosion resistance of the weld metal, particularly the erosion corrosion resistance.[3] For this reason, it is common to weld 90/10 copper-nickel with 70/30 copper-nickel filler. The higher nickel content greatly increases the iron solubility limit and there is no risk of iron precipitation on cooling. Note that slow cooling of thick section welds may cause iron to precipitate out of solution in the HAZ. This can be avoided by suitable joint design and the selection of welding parameters when qualifying the welding procedure. Thin sections of 90/10 copper-nickel have been welded with matching filler, because the weld is small and cools sufficiently quickly to keep the iron in solution.

The 70/30 copper-nickel alloy, C71500, is usually welded with a matching composition filler. This produces welds that are not anodic to the parent metal so they will not corrode preferentially. Alloy C71500 has been welded with alloy 400 filler, but this is generally unnecessary because it is more expensive than matching filler and does not generally confer any significant corrosion benefits.

Little et al. examined the corrosion of welded copper-nickel alloys in seawater polluted by sulfides. They concluded that any corrosion would occur, not on the weld, but downstream of a girth weld, on the HAZ.[4]

The welding of the aluminum bronzes is somewhat more complex, particularly for the multi-phase alloys. The subject is discussed in detail by Meigh and also by Spiller.[5-6] Here, a brief overview is presented and readers wishing more information should consult these references.

One problem with welding the aluminum bronzes is the ductility dip. The alloys suffer a loss of ductility around 500 °C and the width of this dip varies with alloy composition,[5] but it can be extensive for some of the aluminum bronzes. This means that cracking may occur on cooling, particularly when the weld is under severe restraint. Meigh gives recommendations for minimizing the risk of cracking, involving both the choice of welding parameters and the selection of the filler composition.[5]

Spiller reports that there is a risk of HAZ cracking when welding aluminum bronzes.[6] However, this has not been seen for many years and it is believed that modern manufacturing methods produce alloys that are no longer susceptible to HAZ cracking.[7]

When welding the single-phase aluminum bronzes, it is usual to use matching composition filler. The alloy containing 8% aluminum and 3% iron is sometimes preferred because iron increases the strength of the weld metal and reduces the risk of cracking.[5] Iron also prevents the formation of gamma 2 phase, which is anodic to the alpha phase and can corrode preferentially.[8]

When welding nickel aluminum bronze (NAB), it is common to use a matching filler. However, the weld bead cools very quickly and not all the beta phase present at high temperature may transform to alpha and kappa phases. Beta phase is highly anodic to alpha phase and will corrode rapidly in seawater. Hence, a postweld heat treatment (PWHT) is desirable to ensure that all the beta phase transforms. The most common heat treatment is that specified by the U.K. Royal Navy for all welds to NAB. This involves heating at 675 °C for six hours followed by air cooling. Where postweld heat treatment is not possible, then the use of Cu/8Al/3Fe filler is an alternative. Although no postweld heat treatment will be required, the joint will not be as strong as one made with matching filler, and may not be as strong as the parent metal. The use of a single-phase filler does not result in any preferential weld corrosion in seawater because the potentials of both aluminum bronzes are similar. Figure 3.2 shows a seam welded NAB pipe that had not been postweld heat treated. After exposure to seawater, the beta phase corroded rapidly with the result that the seam weld "unzipped."

FIGURE 3.2 Corrosion of the beta phase in an NAB seam welded pipe that had not been post weld heat treated.

Provided that they have been heat treated, NAB welds do not generally cause a corrosion problem in seawater. However, the Royal Navy has seen corrosion of NAB, both welded and unwelded when it is exposed to slowly flowing seawater, or alternating stagnant and flowing conditions for extended periods. The work of Maselkowski et al. showed the most pronounced attack in the HAZ of NAB welds with the corrosion taking the form of selective attack of the kappa 3 phase (the structure of NAB is discussed in Chapter 2).[9] Because this phase is in long thin stringers, the attack can be very deep. However, if the NAB is put into flowing seawater such that a protective film forms, then no selective phase attack occurs subsequently. This explains why NAB welds give few problems in industrial service, where there is pressure to get equipment into service rapidly.

3.4 Stainless Steels

The super ferritic stainless steels (Cr > 25%) are not usually welded because of the risk of cracking, particularly in thicker sections. Martensitic stainless steels may be welded with roughly matching composition consumables, but require a full PWHT to restore the strength and ductility. Neither class of alloy has any significant use in desalination plants.

3.4.1 Austenitic Stainless Steels

The low-alloy austenitic stainless steels, such as 304L and 316L, are usually welded with matching filler metal by any of the arc welding processes. If there is a concern about delta ferrite in the weld metal, then a higher nickel filler can be used to ensure a fully austenitic weld metal. Such an alloy is 309 (UNS W30913 or W30920), which are nominally Fe/23Cr/12Ni and Fe/23Cr/12Ni/2Mo, respectively. The 309 filler is often used to join 300 series stainless steels to carbon steel to avoid excessive delta ferrite formation caused by dilution.

One problem that can affect austenitic stainless steel welds is solidification cracking. This issue is discussed in detail by Gooch: he concludes that filler metals that produce 5–10% delta ferrite largely eliminate the risk of solidification cracking in well-made welds.[10] Even with fully austenitic weld metals, the risk of cracking can be drastically reduced by using a low travel speed during welding.[10]

Higher alloy austenitic alloys, such as 904L and the 6% Mo alloys, suffer from dendritic segregation when welded with matching composition filler. The dendrites in the weld metal are depleted in chromium and molybdenum, greatly reducing the corrosion resistance. Hence, it has become standard practice to weld these alloys with a nickel-based consumable, such as alloy 625, where the chromium and molybdenum content in the dendrites is still acceptable for most applications.

One feature of welding with fillers such as alloy 625, is the creation of an unmixed zone at the weld root. This is where the parent metal has melted and then refrozen without mixing with the filler metal. This area has been shown to have highly depleted dendrites and a much lower corrosion resistance. By careful choice of joint design and welding parameters to ensure better mixing of the weld pool, it is possible to minimize the unmixed zone, although it is difficult to prevent completely.[11] The use of a higher alloy filler has no effect on unmixed zone formation.

When stainless steels are being used beyond their limits of use, it is common to weld overlay with a more corrosion resistant alloy, and alloy 625 has been widely used for this in seawater. When the component is creviced, such as a flange face, then it has been demonstrated that the crevice corrosion resistance of as-welded 625 is not as good as that of the parent 6% Mo alloy, because of the dendritic as-welded structure.[12] Rogne et al. conducted crevice corrosion tests on 6% Mo stainless steel flanges with different weld overlays and two different gasket materials in seawater at +600 mV SCE, to simulate chlorination (see Chapter 4).[12] The results in Figure 3.3 show that the aramid fiber gasket created a much less aggressive crevice than the polytetrafluoroethylene (PTFE) gasket. They also showed that a weld overlay with alloy 625 had inferior crevice corrosion resistance to that of the parent metal. To get an increase in the crevice corrosion resistance, it was necessary to use a higher alloyed nickel-base filler such as alloy 59.[12]

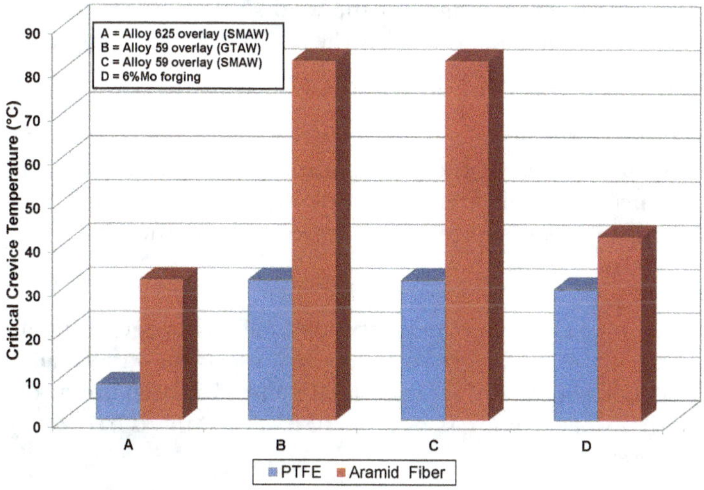

FIGURE 3.3 Effect of weld overlay composition and gasket on crevice corrosion of 6% Mo austenitic stainless steel in seawater at +600 mV SCE.[12]

The necessity for tests of austenitic stainless steel welds—over and above the requirements of whatever welding standard is being used (e.g., ASME's

"Boiler and Pressure Vessel Code Section IX–Welding, Brazing and Fusing Qualifications")–is discussed in Chapter 8.

3.4.2 Duplex Stainless Steels

Autogenous welding of duplex stainless steels is not generally recommended because the weld metal will be highly ferritic because of the fast cooling rate. However, if the weld is solution annealed after welding, the microstructure will be restored to a normal phase balance with good properties. Without postweld heat treatment of autogenous welds, there will be a significant reduction in corrosion resistance and toughness.

If duplex alloys are welded with a matching composition filler, then the weld will be ferritic, as with autogenous welding. However, a full, postweld solution anneal heat treatment will completely restore the microstructure and give good properties.

However, most welding of duplex stainless steels requires their being used in the as-welded condition. The filler wires for this requirement are usually matching in composition with an extra 2–3% nickel. Common examples are 2209, 2509, and Z100X. The extra nickel enables sufficient austenite to form during the fast cooling of the weld so that the 50/50 austenite/ferrite balance is maintained. With the superduplex alloys, ASME rules now permit welding of 2507 and Z100 with either of the two alloys' fillers.

However the weld is made and whatever filler is used, it is most important that the final weld has a good phase balance, in the range 40–60% ferrite for preference, although some standards allow a phase balance up to 65% ferrite. This will ensure both good mechanical properties and good corrosion resistance.

Manufacturers of alloys and welding products often have guidelines for joint design, heat inputs, and interpass temperatures for the various alloys, e.g., Ref. 13. A good overview of welding duplex alloys has been produced by the International Molybdenum Association (IMOA) and this is free to download from their website.[14]

The root pass of a duplex weld can sometimes be a little ferritic, but reheating during the second pass of a multipass weld will cause further transformation of ferrite to austenite. This usually ensures a good phase balance in the root, which is the side of the weld normally in contact with corrosive fluids. The capping pass is not reheated and may sometimes be a little ferritic, but this is not usually exposed to corrosive fluids. If it is, the corrosion resistance can be improved by pickling (see below). Figure 3.4A shows a typical root microstructure of a multipass weld, while Figure 3.4B shows a slightly ferritic capping pass. The ferrite is the darker phase.

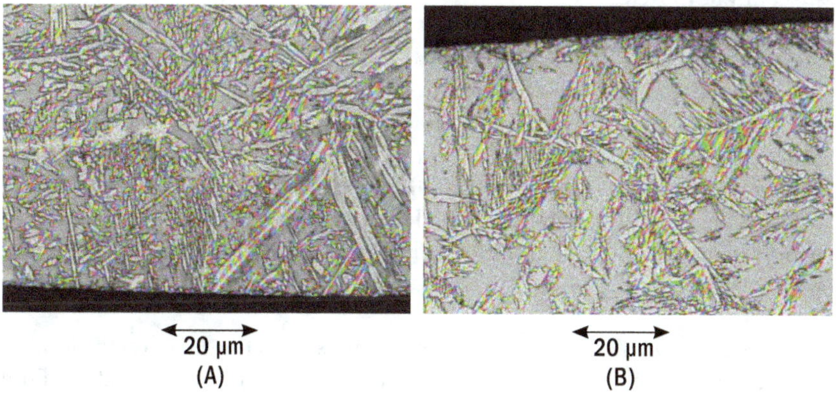

FIGURE 3.4 Typical microstructures of duplex stainless steel multipass welds: (A) weld root, B) weld cap.

It is often stated that welds are not as corrosion resistant as parent metal, but if correctly made, duplex welds are often similar in performance. Walker and Gooch concluded that a filler wire with a PREN (see Chapter 2) about 2 points higher than that of the parent metal (2205) would give welds with similar corrosion resistance.[15] Typical commercial welding fillers for 2205 usually meet this requirement, particularly if this is requested, because of the abundant use of 2205 by industry.

Where concern exists about the corrosion resistance of the weld or the correct filler is not available, it is common to weld with a higher alloy duplex filler. Thus, lean duplex alloys are often welded with 2209 filler, and 2205 has been welded with superduplex filler. For the superduplex alloys it would be necessary to consider a nickel alloy filler for higher corrosion resistance. Alloy 625 (UNS N06625) is *not* suitable because segregation in the weld metal reduces its corrosion resistance, as discussed above for super austenitic alloys.

Superduplex has been successfully welded with C-276 (UNS N10276), C-22 (UNS N06022) and alloy 686 (UNS N06686) fillers, which have high chromium and molybdenum contents. Note that these nickel alloy fillers may not match the strength of the superduplex parent metal.

Gooch comments that the threshold stress for SCC of duplex welds is no lower than that for parent metal; and the fact that weld metal is usually somewhat stronger than parent metal means that welds are not particularly susceptible to SCC.[16]

Sedriks comments that welds may be more susceptible to SCC if they are poorly welded such that third phases, such as sigma, are precipitated, or the ferrite content is particularly high.[17] The latter caused the SCC of an offshore vessel made from 2205 when seawater was trapped under highly absorbent insulation at around 100 °C.[18] The SCC resistance of duplex welds is thoroughly reviewed by McIntyre and Francis and they show that properly made welds are not particularly susceptible to cracking compared with parent metal.[19]

It is well known that if superduplex stainless steel is incorrectly welded, sigma and/or chi phase may precipitate in the HAZ and the chromium- and molybdenum-depleted zone around the particles is then susceptible to corrosion. Francis and Warburton modeled the depleted zone and calculated the critical sigma content for corrosion to occur in warm, chlorinated seawater.[20] The calculation showed that for 2 μm diameter sigma particles the threshold concentration was ~ 4%, and this was confirmed by exposure tests of Z100 welds with different sigma contents. The threshold was very susceptible to the third phase particle diameter and fitness for purpose tests were recommended to determine whether a particular weld qualification was acceptable. The ASTM G48 test at 35 °C has been recommended as a suitable qualification test, with a pass being no pitting and a weight loss < 4 g/m^2, e.g., Ref. 21. The NORSOK standard for welding, M-601, also requires a G48 test temperature of 35 °C in the 2015 revision.[22] The whole issue of appropriate tests for duplex welds is discussed in Chapter 8.

The author has carried out a lot of corrosion tests of welded duplex stainless steels and generally welds are as corrosion resistant as parent metal in many fluids. This is confirmed by Gunn for some common fluids.[23] There are numerous publications showing that cleaning welds improves their corrosion resistance compared with the as-welded condition. Gunn shows some data for both 2205 and superduplex welds, including manual wire brushing, grinding, blasting, and pickling with pastes or gels.[23] The results clearly show pickling to give the greatest increase in the critical pitting temperature (CPT).

Figure 3.5 shows the CPT of various welds in synthetic seawater at a potential of +600 mV SCE, typical of the potential of stainless steel in chlorinated seawater. The tests were with 2205 welded with either the regular 2209 filler or with superduplex filler Z100X, and Z100 welded with its Z100X filler. The pickling was carried out with commercial paste based on nitric/hydrofluoric acids. The results show that the CPT of all the welds can be improved by pickling, although the least advantage was gained by pickling 2205 welded with superduplex filler.

It has been pointed out that when duplex stainless steels are welded with pure argon backing gas and all the air is flushed out, it is possible to lose nitrogen from the weld root, which reduces corrosion resistance.[24] Several authors have carried out tests using shielding and backing gases containing nitrogen and some of this data is reviewed by Gunn.[23] The results show significant increases in the CPT when nitrogen is present in the shielding gas for both 2205 and superduplex. There was a greater increase in CPT for superduplex with nitrogen in the shielding gas. Byrne et al. showed that the biggest improvement in weld corrosion resistance was obtained with argon with 2% nitrogen as the shielding gas, and nitrogen plus 10% hydrogen as the backing gas.[24] Figure 3.6 shows the effect of gas composition on the CPT in the ASTM G48 ferric chloride test for both 2205 and Z100 superduplex.[24] The beneficial effect of nitrogen in the welding gas on the pitting resistance of both alloys is clearly shown.

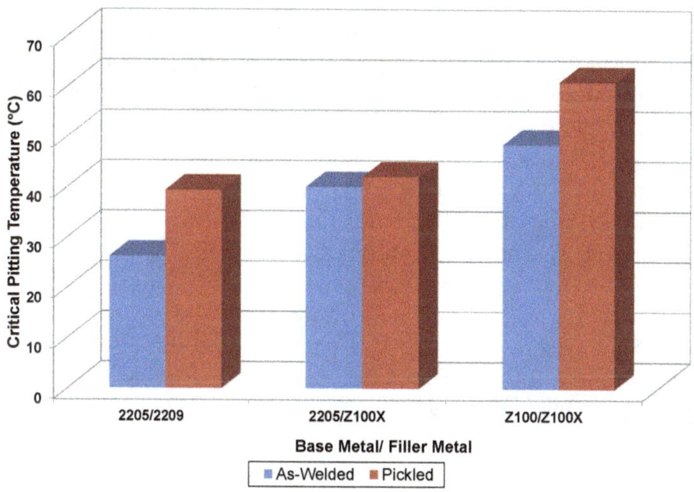

FIGURE 3.5 Effect of pickling and filler metal on CPT in seawater at +600 mV SCE.[23]

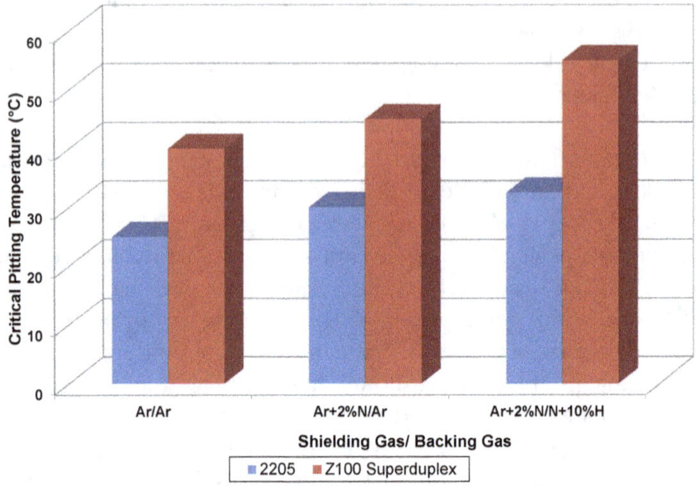

FIGURE 3.6 Effect of welding gas composition on CPT in ASTM G48E test.[23]

It has been shown by Francis and Byrne that the resistance of superduplex stainless steel welds to pitting in seawater (and the crevice corrosion resistance of parent metal) can be improved by a "soft" startup.[25] A typical start-up regime follows:

- Start in cold seawater for two days (minimum)
- Run in cold chlorinated seawater for 5 days (minimum)
- Turn on heat exchangers and operate normally.

Francis and Byrne give an example where a soft startup prevented pitting of welds and crevice corrosion of parent metal for some superduplex piping.[25] The piping was downstream of a gas cooler running at high temperatures and the discharge seawater temperature exceeded 50 °C.

Additional testing and QA/QC requirements, over and above those in ASME to obtain good quality duplex stainless steel welds, are discussed in detail in Chapter 8.

3.5 Nickel Alloys

Nickel alloys can be welded by all of the common arc welding processes, but the selection of filler metal requires careful consideration to ensure that the weld metal is cathodic to the parent metal and it has adequate corrosion resistance. All of the nickel alloys suffer segregation on welding, which reduces the corrosion resistance.[26] This is discussed below for specific alloys.

Nickel-copper alloys are welded with a filler metal of similar composition with the addition of small amounts of other elements to combine with oxygen and nitrogen from the atmosphere to avoid porosity in the weld. Welds in K-500 are not as strong as the parent metal and postweld heat treatment is required to maintain the strength. A Ni-Cr-Fe alloy can also be considered where strength is important[27] and postweld heat treatment is not possible.

In an examination of the corrosion of welded alloy 400, Lee et al. carried out tests in seawater containing inorganic sulfide or sulfate reducing bacteria (SRB) cultures.[28] The welds used a filler that was similar to parent metal, but with an increased silicon and manganese content plus ~ 2% titanium (UNS N04060). In the inorganic sulfide, there was more or less uniform corrosion with no pitting anywhere, even after 20 weeks exposure. Samples exposed to seawater containing SRB's had suffered pitting of the parent metal after 12 weeks exposure. After 14 weeks exposure, there was attack at the weld fusion line, but no attack of the weld metal after 20 weeks. This shows that the welds are no more susceptible to localized attack in this environment compared with parent metal.

The segregation problem discussed above for austenitic stainless steels also occurs with the Ni-Cr-Mo alloys. The ones particularly affected are alloys 825 and 925, and it is customary to weld these with alloy 625 and alloy 725 consumables, respectively.

The higher alloyed Ni-Cr-Mo alloys are welded with matching composition filler, sometimes with small additions of other elements to improve weldability. The reduced crevice corrosion resistance of alloy 625 in seawater, when used as a weld overlay was discussed above in the section on austenitic stainless steels, and the alternative alloys are C-22, 59, C-2000, and 686.

3.6 Titanium

Unlike stainless steels, titanium alloy components are generally highly insensitive to variations in surface condition in typical marine/seawater exposures. As such, no corrosion concerns exist for titanium components in the pickled, ground, machined, peened, grit-blasted, or cold- or plastically worked surface conditions. Although typically not a problem encountered with qualified fabrication practice, smeared-in (embedded) surface iron or carbon steel is only a potential localized corrosion concern in hot (> 80 °C) seawater service with unalloyed titanium (grades 1, 2, and 3) and Ti-6Al-4V.[27] The more crevice corrosion resistant grades, such as the Pd- or Ru-enhanced, and the Mo-containing alloys, are generally resistant to this rare higher-temperature seawater phenomenon. Other surface films, coatings, and/or residues including heat-tint (thermal oxide films), coatings, organic residues, and biofilms also have no detrimental influence on titanium's seawater resistance.

Although titanium mill products are normally surface-conditioned and are free of any surface alpha case (i.e., a brittle, diffused-in oxygen layer), any subsequent fabrication steps in air involving temperatures exceeding approximately 540 °C, for example, hot forming in air or postweld stress-relief annealing, should consider whether oxide scale and alpha case removal is necessary. Generally, the depth of alpha case formed is a function of time at temperature, and is generally not a concern for most applications when it is less than 3–5 μm deep. Total surface alpha case removal should be done in dynamic, cyclically loaded components in which S-N fatigue life is critical, or with high-strength alloy structural components.[27]

The titanium alloys commonly used in marine service are routinely welded using conventional fusion methods, such as gas tungsten arc welding (GTAW), plasma, laser, and electron beam (EB) welding. The two key aspects for successful titanium welding are avoiding contamination by ensuring

1. adequate inert gas shielding of face and root sides for molten and heat-affected metal surfaces above approximately 26 °C, and
2. good joint surface cleanliness prior to welding. Guidelines for welding titanium are well established and can be readily accomplished in most stainless steel welding shops with minor adjustments and training.[29–31]

This means that titanium can be easily welded in the fabrication shop by experienced welders, but it is more difficult to weld in the field than stainless steels, because of the strict requirements on cleanliness and freedom from oxygen by good inert gas purging.

Other than achieving contamination-free, ductile welds, titanium welding is a fairly robust process and is tolerant to minor variations in welding parameters to achieve acceptable properties. Although postweld stress-relief annealing is normally unnecessary for lower-strength titanium grades with yield strengths below approximately 585 MPa, it is recommended for higher-strength alloys typically used in structural or dynamically loaded applications.[27]

3.7 Polymers

As with all engineering materials, the fabrication method dictates the performance in the end-use application. Fiberglass reinforced plastic (FRP) equipment and piping is typically round because it is built on moulds or mandrels. The work is fabricated from the inside out by first applying the corrosion liner then applying the structural part of the laminate. This is commonly applied by computer-controlled equipment, which applies the glass fibers and resin at the same time. As the mandrel turns, the glass fibers are wound on at a predetermined angle. The resin and catalyst are premixed and applied at the same time, either by running the fibers through a bath of the catalyzed resin or by spraying it onto the mandrel as the glass is wound on. Chopped fibers may be used as part of the corrosion liner and this is typically applied by computer-controlled equipment to ensure a uniform thickness of laminate for the entire corrosion barrier. Most FRP equipment is designed and fabricated in accordance with ASME RTP-1.[32] This widely used standard includes quality and testing requirements. Piping is made to a number of different standards. In general, all FRP piping and equipment is made with a 10:1 safety factor (i.e., stress-to-fracture ratio compared with the maximum rated stress). Only the structural layer of the laminate is included in the strength calculations, the corrosion/abrasion liner and external fire retardant and UV inhibitive layers are excluded from any mechanical calculations.

FRP can be fabricated in the workshop and transported to sites, but as plants have become larger, the demand for very large FRP structures has increased. There are multiple ways to field erect tanks and process vessels on site, but the two most common are oblation and in-place filament winding. For a description of these, refer to Ref. 21. Smaller tanks are made in the shop and are shipped to sites for installation.

FRP pipe is made to a range of standards including ASTM D2996[33] for filament wound pipe and ASTM D6041[34] for contact-molded FRP pipe and fittings. There is no universally accepted standard for FRP pipe design, but an ASME working group is writing an equivalent to the ASME B31.3 standard that is used for metallic pipe. The absence of a universally accepted piping standard means that individual FRP piping system producers have their own set of standards based on ASTM, British Standards, or other national standards. The end user must be clear which standard their FRP piping standard is fabricated to, and that the piping standard is appropriate for the intended duty.

Like all materials, FRP can give problems if it is not correctly joined. Strandmyr reported leaks of FRP seawater piping at joints due to poor workmanship.[35] Joints in FRP are generally glued and personnel doing this do not generally have the rigorous training and approval that is required by welders fabricating metal piping joints. Hence, the quality of joints depends on what level of QA/QC is being applied by the specific project. A similar problem was encountered by the author's company at joints on an NPS 36 seawater pipeline for a Middle East multistage flash distillation (MSF) desalination plant. Stannard reported problems offshore in ensuring the integrity of large diameter FRP pipe joints.[36] He said that water hammer had caused failure of joints that had passed a hydrostatic pressure test in systems subject to fluctuating pressures.

There are alternative methods of joining FRP pipe, particularly when it is large diameter, mostly in long straight runs. The use and joining of FRP for seawater piping on Shell's tanker fleet was extensively studied and implemented by Grimm.[37] Grimm makes the following recommendation: Glued joints should be avoided and all joints should be of the socket and spigot type with double (or triple for very large diameters) 'O' ring seals. These joints are easy to make and can be pressure tested immediately after making and before filling the pipes with water. If a joint leaks it can be undone, the seal changed, and the joint remade in less than 30 minutes. Not only are the joints quick to make and test, but the local work force can be quickly trained in the necessary skills.

With FRP pipe, flanged joints are required to connect to other equipment (e.g., valves) and the flanges are usually glued to the pipes. 'O' ring joints can be used on long pipe runs, but would require locking devices or concrete thrust blocks (if buried) to resist axial forces.

REFERENCES

1. Bond, S., "Corrosion of Welded Components in Marine Environments," Prevention and Management of Marine Corrosion Conference (Cambridge, United Kingdom: TWI Ltd., 2003).
2. Francis, R., *Galvanic Corrosion: A Practical Guide for Engineers*, 2nd ed. (Houston, TX: NACE International, 2017).
3. Francis, R., *The Corrosion of Copper and Its Alloys; A Practical Guide for Engineers* (Houston, TX: NACE International, 2010).
4. Little, B.J., P.A. Wagner, and J. Jacobus, "The Impact of Sulfate Reducing Bacteria on Welded Copper-Nickel Seawater Piping Systems," CORROSION/88, paper no. 81 (Houston, TX: NACE International, 1988).

5. Meigh, H.J., *Cast and Wrought Aluminium Bronzes–Properties, Processes and Structures* (London, United Kingdom: Institute of Materials, Minerals and Mining, 2000).
6. Spiller, K.R., "Practical Aspects of Welding Aluminium Bronze," *Welding and Metal Fabrication* (November 1971): p. 401.
7. Gittos, M.F., TWI (Cambridge, United Kingdom), Private Communication with R. Francis (2009).
8. Campbell, H.S., "Aluminium Bronze Alloys Corrosion Resistance Guide," Publication No. 80 (Potters Bar, United Kingdom: Copper Development Association, 1981).
9. Maselkowski, C.R., and R. Francis, "The Effects of Chlorine on Materials for Seawater Cooling Systems: A Comparison of Specimens from Recirculating and Once-Through Environments," *BNF Metals Technology Centre Research Report, R506 (Supplementary Report)*, Internal report to BNF MTC membership, July 1985 (available from R. Francis).
10. Gooch, T.G., "Solidification Cracking of Austenitic Stainless Steel Welds," ASM Materials Week (Materials Park, OH: ASM International, 1990).
11. Gooch, T.G., "The Unmixed Zone in CRA Weldments, An Old Effect Re-emerges," *Stainless Steel World* 7, 9 (1995): pp. 29–35.
12. Rogne, T., J.M. Drugli, T. Solem, H. Salbu, and H. Skjellevik, "Crevice Corrosion Properties of Weld Overlays of Ni-Based Alloys Compared to 6%Mo Steels for Seawater Applications," CORROSION 1998, paper no. 696 (Houston, TX: NACE International, 1998).
13. Rolled Alloys, "Welding Guidelines for ZERON 100," Publication No. 105, 2009, https://www.rolledalloys.com/shared-content/technical-resources/fabrication-information/ZERON-100_FI_US_EN.pdf (December 2018).
14. IMOA, *Practical Guidelines for the Fabrication of Duplex Stainless Steels*, 3rd ed., https://www.imoa.info/download_files/stainless-steel/Duplex_Stainless_Steel_3rd_Ed.n.pdf (London, United Kingdom: International Molybdenum Association, 2014).
15. Walker, R.A., and T. G. Gooch, "Pitting Resistance of Weld Metal for 22Cr-5Ni Ferritic-Austenitic Stainless Steels," *British Corrosion Journal* 26, 1 (1991): p. 51.
16. Gooch, T.G., "Corrosion Resistance of Welds in Duplex Stainless Steel," Duplex Stainless Steels '91 (Les Ulis Cedex A, France: Les editions de physique, 1991): p. 325.
17. Sedriks, J., *Corrosion of Stainless Steels*, 2nd ed. (New York, NY: J. Wiley and Sons, 1996).
18. Oystetun, I., K. Johansson, and O.B. Andersen, "Stress Corrosion Cracking of Duplex SS, UNS S31803, on the Gyda HP Separator," Offshore Technology Conference, paper no. 7207 (Houston, TX: Offshore Technology Conference, 1993).

19. McIntyre, D.R., and R. Francis (eds.), "Environmental Cracking," MTI publication 15 (St Louis, MO: Materials Technology Institute, 2016).
20. Francis, R., and G.R. Warburton, "A Model for the Corrosion of the Depleted Zones Around Sigma Precipitates Produced During Welding of Superduplex Stainless Steel," Stainless Steel World Conference 1999, paper no. 99006 (Zutphen, Netherlands: KCI Publishing: 1999).
21. Clarke, S., and R. Francis, *The Selection of Materials for Mineral Processing Operations* (Houston, TX: NACE International, 2016).
22. NORSOK M-601, Rev. 06, "Welding and Inspection of Piping" (Oslo, Norway: Standards Norway, 2015).
23. Gunn, R., (ed.), *Duplex Stainless Steels; Microstructure, Properties and Applications* (Cambridge, UK: Abington Publishing, 1997).
24. Byrne, G., G. Warburton, J. Wilson, and R. Francis, "Fabrication of Superduplex Stainless Steel for Optimum Seawater Corrosion Resistance," World Congress on Desalination and Water Reuse (Topsfield, MA: International Desalination Association, 2011).
25. Francis, R., and G. Byrne, "Experiences with Superduplex Stainless Steel in Seawater," *Stainless Steel World* 16 (2004): p. 53.
26. McCoy, S., L. Shoemaker, and J. Crum, "Corrosion Performance and Fabricability of the New Generation of Highly Corrosion Resistant Nickel-Chromium-Molybdenum Alloys," Stainless Steel World Conference (Zutphen, The Netherlands: KCI Publishing, 2001),
27. Powell, C., and R. Francis (eds.), "The Corrosion Performance of Metals for the Marine Environment: A Basic Guide," EFC Publication No. 63 (London, United Kingdom: Institute of Materials, Minerals & Mining, 2012).
28. Lee, J.S., K.L. Lowe, R.I. Ray, and B.J. Little, "Corrosion Mechanisms of UNS N04400 in Seawater," CORROSION 2003, paper no. 214 (Houston, TX: NACE International, 2003).
29. Donachie, M.J., Jr., (ed.), *Titanium: A Technical Guide*, 2nd ed. (Materials Park, OH: ASM International, 2000).
30. Leyens, C., and M. Peters, (eds.), *Titanium and Titanium Alloys-Fundamentals and Applications* (Weinheim, Germany: Wiley-VCH GmbH & Co., 2003).
31. Nagler, H., D.F. Hasson, and C.S. Young, "Arc Welding of Reactive Metals and Refractory Metals," *Metals Handbook,* 9th ed., Vol. 6 (Materials Park, OH: ASM): pp. 446–456.
32. ASME RTP-1, "Reinforced Thermoset Plastic Corrosion-Resistant Equipment" (New York, NY: American Society of Mechanical Engineers, 2011).
33. ASTM D2996 (latest revision), "Standard Specification for Filament Wound 'Fiberglass' (Glass-Fiber-Reinforced Thermosetting-Resin) Pipe" (West Conshohocken, PA: ASTM International).

34. ASTM D6041 (latest revision), "Standard Specification for Contact-Molded "Fiberglass" (Glass-Fiber-Reinforced Thermosetting-Resin) Corrosion Resistant Pipe and Fittings" (West Conshohocken, PA: ASTM International).
35. Strandmyr, O., "Seawater Materials–The Norwegian Experience," Corrosion and Materials Offshore Conference (Oslo, Norway: Norwegian Society of Engineers and Technologists, 1999).
36. Stannard, D., Presentation to the Marine Corrosion Forum (London, United Kingdom, January 2005).
37. Grimm, G.C., and A. Bliall, "The Use of GRP Piping in the Oil and Gas Ballast Water System of the Draugen Gravity-Based Structure," Conference on Polymers in the Marine Environment (London, United Kingdom: Institute of Marine Engineers, 1991).

4 Types of Corrosion

This chapter discusses the main factors that affect corrosion in all types of desalination plants. Corrosion is discussed for each alloy group separately, because the different alloys can suffer corrosion from different mechanisms in desalination environments.

4.1 Carbon Steel

Carbon steel is a low cost construction material for seawater plants, but it is important to recognize that its corrosion rate varies significantly when used as piling. Figure 4.1 shows relative corrosion rates at different elevations in seawater. The highest corrosion rate when fully submerged is just below the mean low-tide level, because this water is usually saturated with dissolved oxygen and water flows are often highest at the surface. The splash zone has the highest rate of attack, because seawater is continuously splashing onto the metal and then running off or evaporating. The relatively thin water films are saturated with oxygen, and evaporation can further concentrate the solution. The tidal zone has a much lower corrosion rate and this is thought to be because it is partially cathodically protected by the more active zones above and below. Corrosion rates below the mud line may be low, but can increase if there is microbial activity (see below).

In quiescent seawater, carbon steel corrodes uniformly at about 0.1 mm/y. The rate of corrosion is a function of both time and temperature, as well as dissolved oxygen. Figure 4.2 shows the data of Melchers for 1 and 5 year exposures at different sites around the world.[1] The corrosion rate can be seen to decrease after 5 years (except at 10 °C), particularly at higher seawater temperatures. This is because the corrosion product layer thickens and there is often heavy fouling, both of which restrict access of fresh aerated seawater to the metal surface. The higher rate of corrosion after 5 years at 10 °C is thought to be due to less fouling and a thinner layer of corrosion products than at higher temperatures. The higher corrosion rate at 10 °C may have been specific to the exposure site that was at 10 °C.

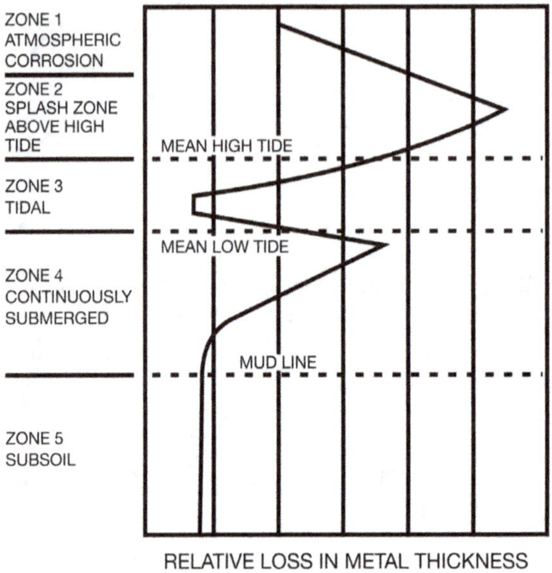

FIGURE 4.1 Relative corrosion rates of carbon steel piling at different locations in the sea.

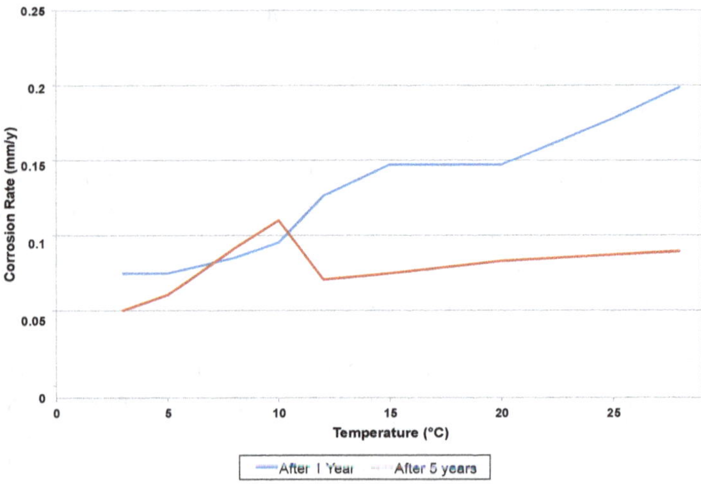

FIGURE 4.2 Corrosion rate of carbon steel in quiescent seawater at different locations around the world.[1]

The corrosion rate of cast iron in seawater is similar to that of carbon steel. However, as the iron is gradually corroded away over a period of years, graphite is left behind. This is known as "graphitization." Graphite is highly cathodic to most metals in seawater (see Section 4.7) and can cause significant galvanic corrosion of some alloys.

The corrosion rate of carbon steel is a function of the seawater flow velocity, as shown in Figure 4.3.[2] At velocities above about 0.5 m/s the corrosion rate becomes significantly high unless other means of corrosion protection are employed.

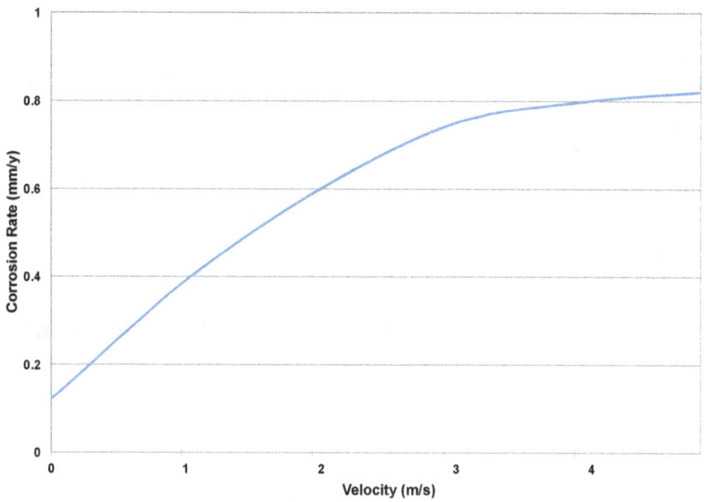

FIGURE 4.3 The effect of velocity on the corrosion of carbon steel in seawater.[2]

In many thermal desalination applications, the dissolved oxygen content is low and this greatly reduces the corrosion rate of carbon steel. A number of models for the uniform corrosion of carbon steel in low oxygen seawater have been proposed, and a widely accepted one is that of Berger and Hau.[3] Figure 4.4 shows the effect of velocity with 20 ppb oxygen at 50 °C for a nominal pipe size (NPS) 6 pipe. At this oxygen content and temperature, the corrosion rate is relatively low even at 7.5 m/s velocity. Figure 4.5 shows the effect of oxygen concentration at 50 °C and 1 m/s for an NPS 6 pipe. This shows that each order of magnitude increase in the oxygen concentration causes roughly an order of magnitude increase in the corrosion rate. The corrosion rates are higher than natural seawater at oxygen concentrations greater than 1 mg/L because there is no fouling to restrict access of seawater to the metal. Figure 4.6 shows the effect of temperature and velocity on corrosion with 20 ppb oxygen for an NPS 6 pipe. At velocities greater than ~ 2 m/s, there are significant increases in corrosion rate as the temperature increases.

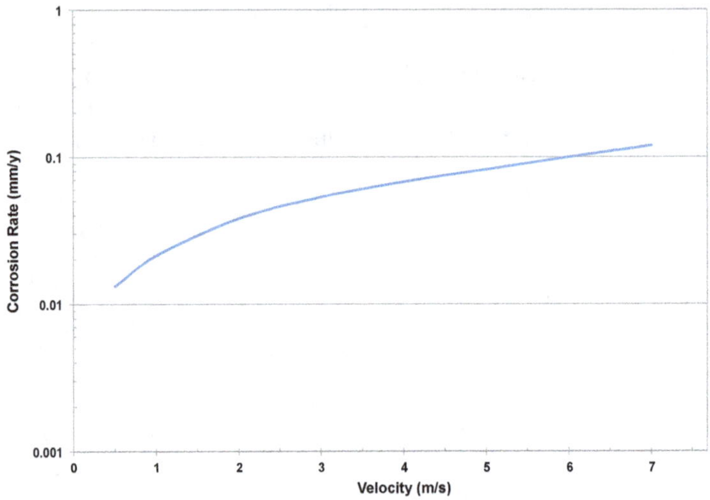

FIGURE 4.4 Effect of velocity on the corrosion of carbon steel in seawater with 20 ppb oxygen at 50 °C for an NPS 6 pipe.

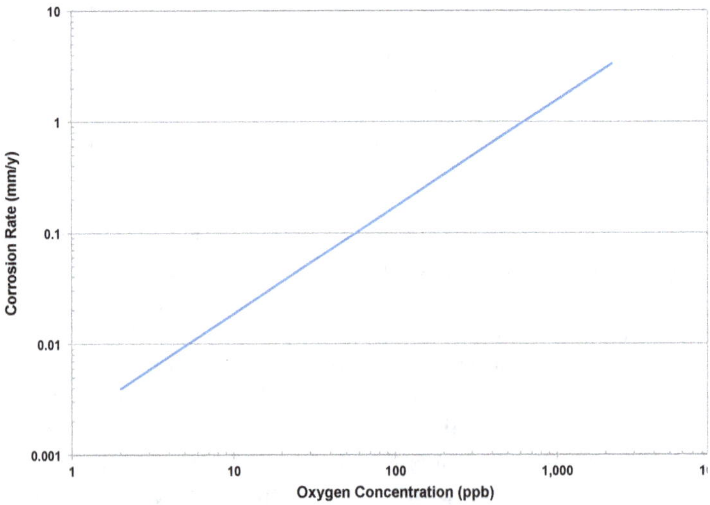

FIGURE 4.5 Corrosion rate versus oxygen concentration for carbon steel at 50 °C at 1 m/s for an NPS 6 pipe.

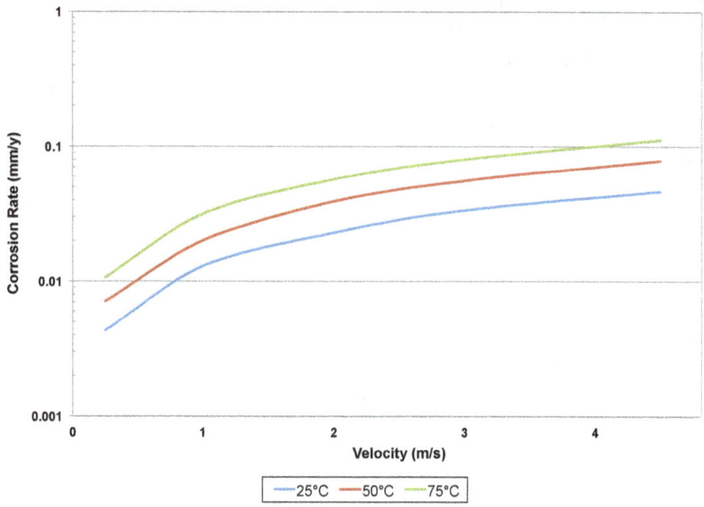

FIGURE 4.6 Effect of velocity and temperature on the corrosion of carbon steel with 20 ppb oxygen and an NPS 6 pipe.

In natural seawater, fouling of steel surfaces occurs readily. Once a significant microbial colony is established on the surface, the aerobic bacteria deplete the local environment in dissolved oxygen, and then the anaerobic bacteria become active. In seawater, the major anaerobic bacteria are the sulfate reducing bacteria (SRB), which reduce sulfate to sulfide as part of their metabolic processes. This sulfide is sufficient to corrode steel and Farinha showed that the corrosion rate of carbon steel in marine muds around the United Kingdom could vary from 15–312 μm/y, depending on the microbial aggressivity of the mud.[4] Under very aggressive conditions, this might increase to ~ 0.5 mm/y. A typical value in a mud from temperate climates is 0.12 mm/y, similar to the corrosion rate in aerated seawater.[5]

In aerated natural seawater, it is not usually uniform corrosion that leads to failure, but pitting.[2] Figure 4.7 shows the maximum pit depth of carbon steel pipe after a 14 month exposure in the Gulf of Mexico.[6] It is clear that the depth of attack is significantly greater than the uniform metal loss after a similar interval. Pitting of carbon steel can also occur under deaerated conditions if SRB are active, producing hydrogen sulfide. Pits up to 0.6 mm deep were found on carbon steel pipe buried for 5 years in a marine mud environment in the United Kingdom.[5]

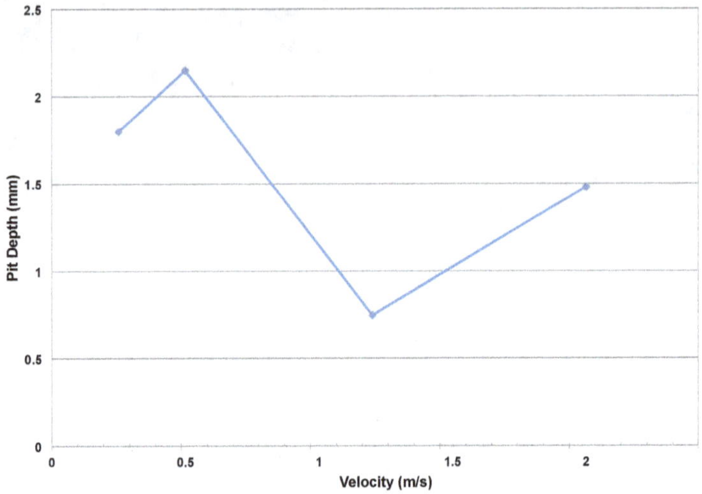

FIGURE 4.7 Effect of velocity on the pitting of carbon steel after 14 months in seawater.[6]

Potential corrosion problems with carbon steel welds were discussed in Chapter 3 and these can largely be overcome by the correct choice of filler and suitable welding parameters.

Because of the potential for high corrosion rates with carbon steel in aerated seawater, it is usual to use coatings and/or cathodic protection to prevent an unacceptable loss of metal.

Carbon steel is widely used in desalination plants for both structural elements and for accessories such as electrical enclosures and equipment housings. Because most desalination plants are by the seaside in warmer climates, the atmosphere is very aggressive. Typically, a marine atmosphere will be "CX," which is the most aggressive, on the scale given in ISO 9223,[7] This means that carbon steel will need a high quality paint coating to protect it for a reasonable time, plus regular inspection and touchups to prevent unacceptable degradation. Such a paint system could be a zinc-rich primer, followed by two coats of epoxy paint, with a polyurethane topcoat for ultraviolet (UV) resistance. It is advisable to consult a coatings expert to determine the optimum coating system and thickness for carbon steel used outside in a particular location. The alternative is to use electrical enclosures of UV resistant fiberglass reinforced plastic (FRP), which is available off the shelf for electrical applications. These will have a long life with minimum maintenance, but will probably need an additional coating in areas with high UV radiation, as discussed in Chapter 2.

4.2 Austenitic Cast Iron

The high nickel content of austenitic cast irons (ACI) plus the spheroidal graphite and chromium addition, gives the alloy good mechanical properties and resistance to corrosion in seawater. Table 4.1 shows the typical corrosion rate of austenitic cast iron compared to carbon steel, nickel aluminum bronze (NAB) and copper-nickel. It can be seen that austenitic cast iron falls between the corrosion rates of carbon steel and NAB.

TABLE 4.1 Typical corrosion rates of some common alloys in quiescent seawater.

ALLOY	CORROSION RATE (mm/y)
Carbon Steel	0.06-0.12
Austenitic Cast Iron	0.04-0.07
Nickel Aluminum Bronze	0.02-0.05
90/10 Copper-Nickel	0.005-0.020

Austenitic cast iron also has very good resistance to erosion corrosion in seawater, better than gray cast iron and admiralty gunmetal, as shown in Table 4.2. This data was generated in natural seawater at 30 °C using a rotating spindle, with samples fastened radially to the spindle, and a tip velocity of 8.2 m/s.[8]

TABLE 4.2 Corrosion rates of some alloys in seawater at 30 °C in a rotating spindle test.[8] (Note that tip velocity = 8.2 m/s.)

ALLOY	CORROSION RATE (mm/y)
Gray Cast Iron	4.47
Austenitic Cast Iron	0.20
Nickel Aluminum Bronze	0.23
Admiralty Gunmetal	1.00

Austenitic cast iron also has a modest resistance to cavitation, which explains its use in seawater pumps, where cavitation can occur under some service conditions. Comparative cavitation data in seawater was not available, but Table 4.3 shows some data in fresh water using the ultrasonic cavitation test method at 20 kHz (ASTM G32).[9-10] It can be seen that the metal loss rate for austenitic cast iron is similar to that of admiralty gunmetal–another alloy widely used for pumps. Although this data is for fresh water, it is not expected that the ranking would change significantly in seawater.

TABLE 4.3 Cavitation metal loss rate in fresh water at 20 kHz.[9]

ALLOY	CAVITATION RATE (mm^3/h)
Ductile Cast Iron	1.3
Austenitic Cast Iron	4.4
Nickel Aluminum Bronze	0.6
Admiralty Gunmetal	4.9

One of the main corrosion problems with austenitic cast iron in seawater is stress corrosion cracking (SCC). In the 1970s and 80s, there were few failures of austenitic cast iron by SCC, and these were largely cured by a stress relief heat treatment. In the 1990s, however, the number of failures increased; subsequent research showed that the stress needed to cause SCC was low, and was initiated when there was corrosion between the matrix and the chromium carbide particles at the surface.[11] Figure 4.8 shows a microsection of a typical SCC crack. Stress relieving extended the time to failure, but did not prevent it.[12] The author believes that the increase in the number of SCC failures was due to economic pressures because the price of nickel increased. Previously, castings had been generously over thick, but as metal costs rose, the wall thickness was reduced, with a resultant increase in local stresses during operation. SCC of austenitic cast iron has not only occurred with pumps, but also with column pipes, as shown in Figure 4.9. Because the corrosion rate of austenitic cast iron is higher in warm waters, the number of SCC failures has been highest in the Middle East. Stress relieving austenitic cast iron castings (450–550 °C) will extend the life in warm seawater, but it will not prevent SCC. The trend has been to replace austenitic cast iron with superduplex stainless steel, which has an excellent track record for pumps, valves, and piping in seawater all over the world.

Although austenitic cast iron contains graphite, it does not generally suffer from graphitization in seawater, because the corrosion rate is so much lower than that of cast iron, as discussed in Section 4.1.

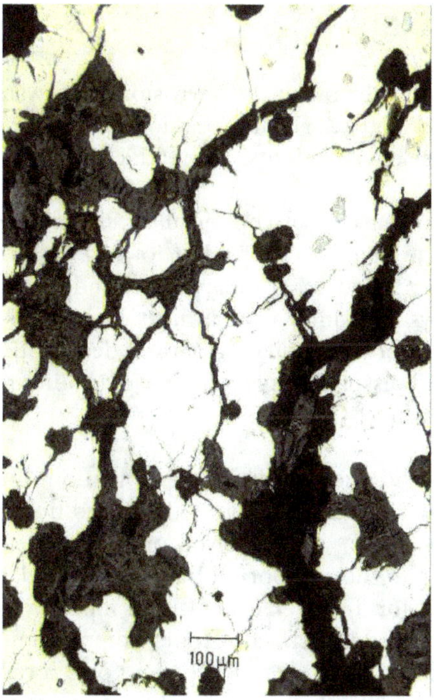

FIGURE 4.8 Microsection of SCC of austenitic cast iron after service in seawater.

FIGURE 4.9 SCC crack in an austenitic cast iron column pipe used with a seawater pump.

4.3 Copper Alloys

This section will only deal with the corrosion in seawater of the copper alloys discussed in Chapter 2. For information on the corrosion of other copper alloys, or other environments, the reader is referred to Reference 13.

The general corrosion rate of copper alloys in quiescent seawater is low, with a typical value being 20 μm/y, but lower rates are found on material that has been exposed for a long period and has formed a protective film. Copper alloys are widely used in heat exchangers in multistage flash (MSF) and multiple-effect distillation (MED) desalination plants and often operate at reduced dissolved oxygen levels, pH values down to 6, and high temperatures up to 110 °C.

Figure 4.10 shows the effect of dissolved oxygen on the corrosion of some copper alloys under typical MSF operating conditions.[14] As expected, the corrosion rate increases with dissolved oxygen content. These corrosion rates are higher than those reported by Oldfield et al.,[15] but those were at lower velocity and for a longer time, where more protective films should form. Ross investigated the effect of water composition, pH, and temperature on the corrosion of copper-nickel alloys over a range of dissolved oxygen concentrations.[16] He found that stable, protective films did not form at pH < 6, even with high oxygen. Corrosion rates decreased with dissolved oxygen concentration. Figure 4.11 shows the corrosion rates of aluminum brass and copper-nickel exposed in an MSF plant in Texas in the United States.[16] The higher oxygen contents were at the lower temperatures and vice versa, and corrosion rates decreased for both alloys at high temperatures. For more details on the corrosion of copper alloys in low oxygen seawater, the reader is referred to References 13–15.

Copper and its alloys can suffer pitting corrosion in seawater, but it is very rare. Francis reports that residual carbon films from the manufacturing process have caused pitting of heat exchanger tubes, but only under conditions of heat transfer.[13] The level of carbon to cause this pitting is high, so the modern internal cleaning methods during the manufacture of heat exchanger tubes should prevent pitting. Test methods to ensure tubes are free of carbon films are discussed in Chapter 8.

Copper alloys can suffer from crevice corrosion in seawater, but it is not like that of stainless steels, because it is caused by a copper ion concentration cell rather than an oxygen concentration cell. It is also different because the anode, where corrosion occurs, is just outside the crevice. In fact, once established, the crevice can be removed and the attack will be self-sustaining. Although crevice corrosion is sometimes found, it is rarely deep (≤0.2 mm), and the author has never seen a copper alloy failure because of crevice corrosion.

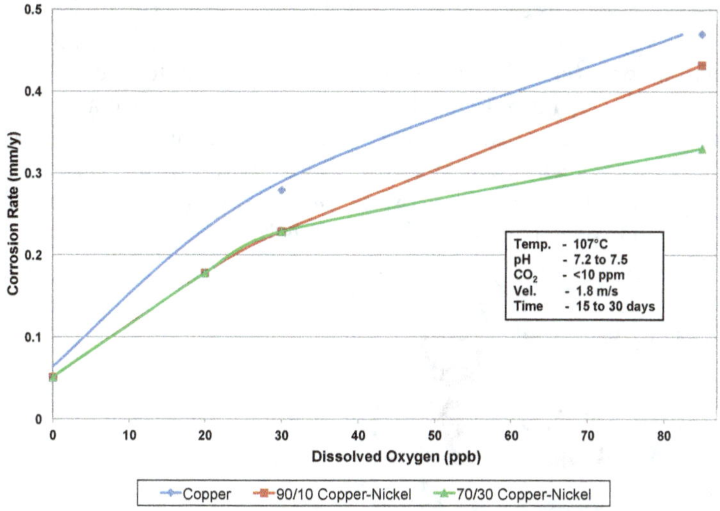

FIGURE 4.10 Effect of dissolved oxygen on corrosion of copper alloys in seawater.[14]

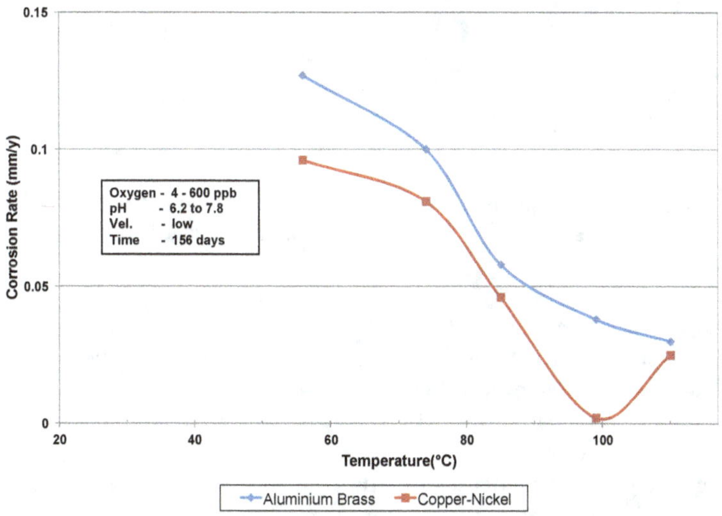

FIGURE 4.11 Effect of temperature on the corrosion of some copper alloys in an MSF plant.[16]

It is well known that the protective film on copper alloys will break down under high velocity or turbulent flow, causing accelerated attack. The severity of attack depends on the alloy composition and the flow velocity. Figure 4.12 shows the range of pit depths for some common heat exchanger alloys in the recirculating jet test at 9.3 m/s.[17] For a discussion of the differences between different erosion corrosion test methods and the effect of recirculating versus once-through seawater see Reference 13. It is clear that the alloys commonly used in thermal desalination plants have good resistance to erosion corrosion. Bog also tested some cast copper alloys using the same technique, and the results are shown in Table 4.4.[18] This shows that the resistance to erosion corrosion increases with tin content, and 10% tin-bronze is nearly as resistant as NAB.

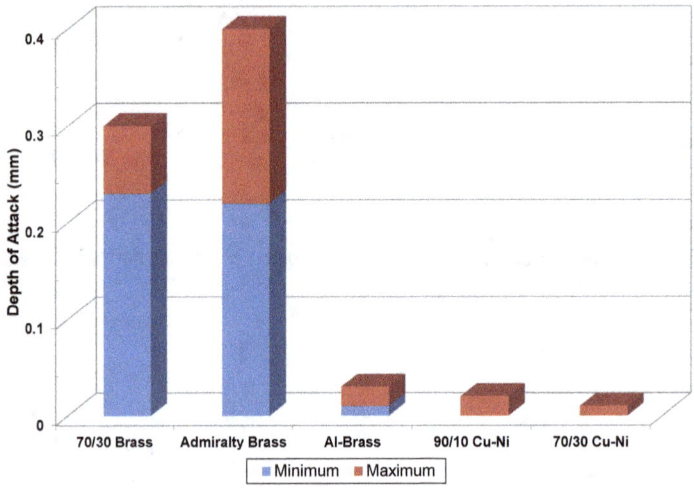

FIGURE 4.12 Depth of attack of some wrought copper alloys in seawater using a recirculated jet test at 9.3 m/s and 18 °C.[17]

Table 4.5 shows some typical design velocities that have been used for copper alloys in clean seawater.[13] Higher velocities are possible in larger diameter pipes (> NPS 3) because there is usually less turbulence than in small diameter pipes and heat exchanger tubes.[13] Note that impeller tip speed is *not* a guide to water velocities in a pump, because the water is also moving and it is relative velocity that is important.[19] In piping systems with sharp bends, the lower velocity in Table 4.5 is preferred. It is more important to keep velocities low in thin walled components such as heat exchanger tubes (typical wall thickness ~ 1.0–1.2 mm) compared with castings for pumps and valves (typical thickness > 15 mm), where a little erosion corrosion can sometimes be tolerated.

TABLE 4.4 Depth of attack in jet impingement tests of some cast alloys in seawater at 9.6 m/s and 20 °C.[18]

ALLOY	UNS No.	MEAN DEPTH (mm)
Mn-Bronze	C86400	0.5
Gunmetal (LG2) (5% Sn)	C83600	0.23
Gunmetal (LG4) (7% Sn)	C92410	0.05
Gunmetal (G1) (10% Sn)	C90500	0.02
Leaded Tin Bronze (3% Pb)	C92900	0.02
NAB	C95800	0.0

TABLE 4.5 Typical velocity limits for copper alloys in clean seawater.[13]

ALLOY	MAXIMUM VELOCITY (m/s)	
	Heat Exchanger	Pipe (> NPS 3)
Admiralty Brass	1.0-1.5	–
Aluminum Brass	1.5-2.0	2.0-2.5
90/10 Copper-Nickel	1.0-3.0	3.0-3.5
70/30 Copper-Nickel	2.5-3.5	3.0-4.0
NAB	–	3.5-4.5

The velocities in Table 4.5 are meant as guidelines and must be used in conjunction with good design practice. The important thing to note is that there is a definite ranking order with copper alloys. The copper-nickel alloys have superior resistance to erosion corrosion compared with the brasses, and NAB has similar resistance to 70/30 copper-nickel. Table 4.5

does not include alloy 66/30/2/2 Cu-Ni-Fe-Mn, which can be ranked as equivalent to or better than NAB. However, this alloy is more often selected when there are suspended solids in the water, or the conditions are more arduous, such as the brine heater in an MSF plant.

The successful use of copper alloys and the avoidance of erosion corrosion depend on the quality of the design. Utilization of the velocities in Table 4.5 alone is not sufficient. It is important to avoid designs that increase local turbulence. Hence, 5-D bends are preferred to 3-D bends, and 3-D bends are preferred to elbows. Traditional 90° elbows produce a lot of turbulence just downstream of the bend, which can lead to erosion corrosion. Smoothing out the flow with a radiused bend minimizes this effect. Similarly, in heat exchangers, it is important not only to use a suitable velocity for the tubes, but also to use a water box design that distributes water evenly to all the tubes without raising turbulence.

Where the water contains entrained silt or sand the erosiveness increases, rising as the solids content increases. Desalination plants rarely have significant suspended solids in the seawater and this is not often a problem. It is discussed at length for copper alloys in Reference 13.

Dealloying can occur with some copper alloys in seawater, the best known of which is dezincification of brasses. The single-phase brasses, such as aluminum brass, are inhibited with arsenic to prevent this. The duplex brasses, such as naval brass and muntz metal, are used for tube sheets, which are usually 35–50 mm thick. Although these tube sheets will dezincify in service, it never gets deep enough to cause a problem. In addition, some heat exchangers have iron or steel anodes in the water box and these will suppress dezincification, as well as inlet end erosion corrosion. Manganese bronze will dezincify quite rapidly in seawater, and failures have occurred when it is used for valve stems. Nickel aluminum bronze provides a corrosion resistant alternative.

It is possible for the copper-nickel alloys to suffer denickelification and the aluminum bronzes to suffer dealuminification, but the conditions under which this can occur are not found in desalination plants. Nickel aluminum bronze can suffer selective attack of any residual beta phase from manufacturing and/or welding, and the avoidance of this by heat treatment is discussed in Chapter 8. The kappa III phase in NAB can be selectively attacked if the alloy is exposed to intermittent use for extended periods so that a protective film does not form. If the NAB is put into service under flowing conditions, a protective film forms in a few weeks and attack of the kappa III phase does not occur.[13]

There are a number of chemicals that can cause SCC of some copper alloys, but these are rarely found in desalination plants, and SCC is not expected under such conditions.[13]

There are also a number of chemicals that may occur in seawater, either as contaminants or as deliberate additions. The main ones are chlorine or hypochlorite (usually in the form of sodium hypochlorite), hydrogen sulfide, and ammonia.

Chlorine or hypochlorite is often added to seawater to control fouling and increase the efficiency of heat exchangers. It is usually added at the seawater intake at a concentration such that by the time it gets to the heat exchangers it is around 0.5 mg/L or less. If the chlorine is added near the pump suction, so that little dilution occurs before contacting metal, it may cause accelerated attack. The accelerated attack is seen as an increase in the susceptibility to erosion corrosion. The author has seen severe erosion corrosion of an NAB pump impeller where chlorine was injected next to the pump suction. Figure 4.13 shows erosion corrosion at a scratch on an aluminum brass tube (dosed with ~ 1 mg/L chlorine and ferrous sulfate), while a scratch on a similar tube with no chlorine additions healed rapidly.

FIGURE 4.13 Erosion corrosion at a scratch on an aluminum brass tube dosed with chlorine (~ 1 mg/L) and ferrous sulfate.

Chlorine can greatly increase the depth of erosion corrosion of heat exchanger alloys, as shown in Figure 4.14. It is clear that aluminum brass is much more susceptible to this form of attack than 66/30/2/2 Cu-Ni-Fe-Mn. and that ferrous sulfate additions to the water can improve the performance, particularly for aluminum brass. Details of how to carry out effective ferrous sulfate dosing are contained in Appendix B. Table 4.6 gives some recommended maximum chlorine doses for the common heat exchanger alloys.[20] Intermittent chlorine dosing is less aggressive than continuous dosing, and all the copper alloys in Table 4.6 can handle 0.5 mg/L chlorine for 1 hour every 12 hours.[20] The alloys with 30% nickel can handle 1 mg/L chlorine at the same dosing interval.

TABLE 4.6 Recommended maximum chlorine dosing for copper alloy heat exchanger tubes.[13]

ALLOY	CHLORINE DOSING LIMIT
All alloys	When ferrous sulfate dosing is being used, switch off continuous chlorination during the dosing period.
Aluminum Brass	Intermittent chlorination preferred (e.g., 1 mg/L for 1 hour every 12 hours, but up to 0.5 mg/L continuous with ferrous sulfate dosing.
90/10 Cu-Ni	Up to 0.5 mg/L continuous
70/30 Cu-Ni	Up to 1.0 mg/L continuous
66/30/2/2 Cu-Ni-Fe-Mn	Up to 2.0 mg/L continuous

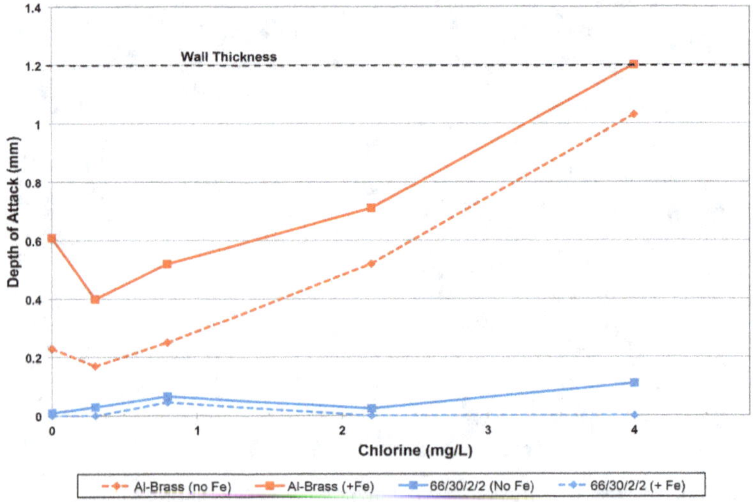

FIGURE 4.14 Depth of attack versus chlorine at 9 m/s for two copper alloys, with and without ferrous sulfate dosing.[20]

Hydrogen sulfide is produced as a metabolic byproduct by SRB in deaerated seawater. This may occur when a system is shut down and left full of seawater or it can occur in bottom mud, when there is mud below the seawater suction/intake. Sulfide films on copper alloys are not protective and the attack becomes more severe with increasing velocity. Hence, failures due to sulfide are most likely in pumps or heat exchangers, but can also occur in partly throttled valves. The most aggressive solution is a mixture of aerated seawater and deaerated seawater with sulfide.[13]

Table 4.7 shows the depth of attack using jet impingement apparatus with recirculating clean seawater and seawater with sulfide additions for some cast copper alloys.[18] It is clear that the ranking in clean seawater and sulfide polluted water is very different. NAB performs very well in clean seawater, but is much more susceptible to attack by sulfide than the tin bronzes and gunmetals. The 10% tin bronze was the most resistant under both clean and sulfide polluted conditions. Also, manganese bronze has low resistance to erosion corrosion under both conditions.

Figure 4.15 shows the effect of increasing sulfide concentration in aerated seawater on the depth of impingement attack at 7 m/s for 90/10 copper-nickel (Figure 4.15A) and 66/30/2/2 copper-nickel-iron-manganese (Figure 4.15B).[21] It is clear that low levels of sulfide had little effect on 90/10 copper-nickel, with somewhat increased attack at higher sulfide concentrations. The 66/30/2/2 Cu-Ni-Fe-Mn alloy was much more susceptible to sulfide attack with the depth of attack increasing rapidly even with low levels of sulfide.

TABLE 4.7 Depth of attack in jet impingement test at 20 °C for some cast copper alloys.[18]

ALLOY	UNS No.	DEPTH OF ATTACK (mm)	
		Clean Seawater	Seawater with Sulfide
Mn-Bronze	C86400	0.5	0.42
Gunmetal (LG2) (5% Sn)	C83600	0.23	0.07
Gunmetal (LG4) (7% Sn)	C92410	0.05	0.04
GUNMETAL (G1) (10% Sn)	C90500	0.02	0.02
LEADED Tin Bronze (10% Sn; 3% Pb)	C92900	0.02	0.03
NAB	C95800*	0.0	0.14

* Aluminum slightly below specification minimum.

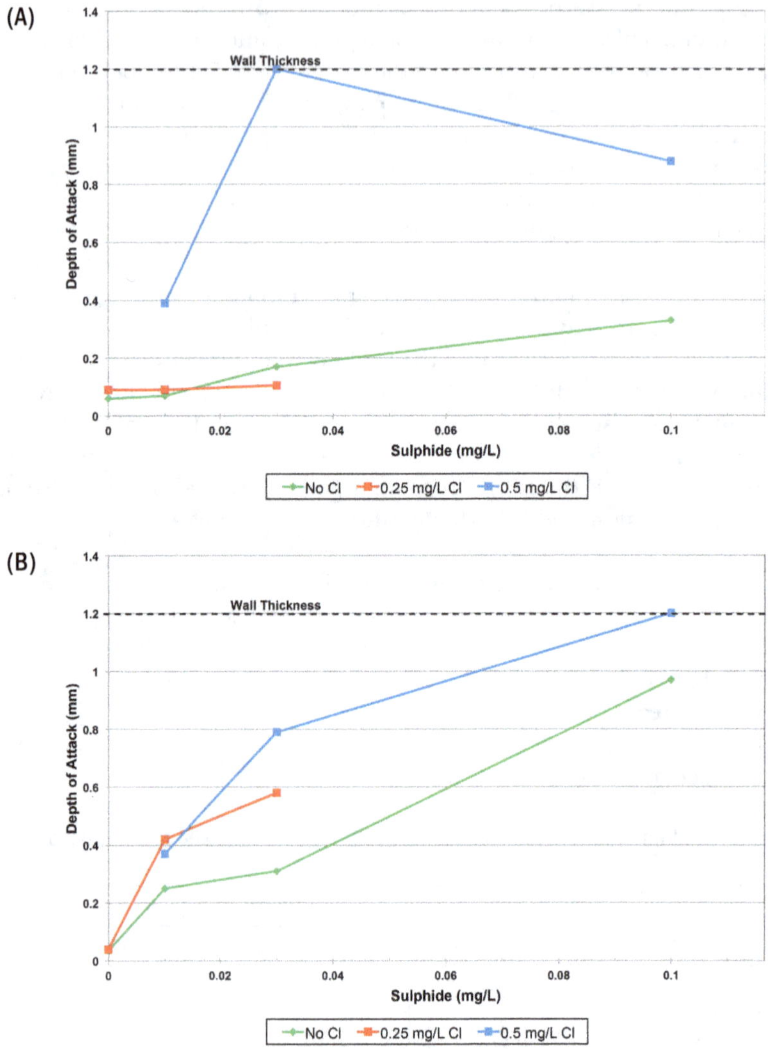

FIGURE 4.15 Effect of sulfide and chlorine on the depth of attack at 7 m/s.[21] (A) 90/10 Cu-Ni, (B) 66/30/2/2 Cu-Ni-Fe-Mn

With low levels of chlorine (0.25 mg/L) as well, the depth of attack on 90/10 copper-nickel was no different, while at 0.5 mg/L chlorine, the depth of attack increased dramatically, resulting in tube perforation in some tests. All levels of chlorine caused a significant increase in the depth of attack of 66/30/2/2 Cu-Ni-Fe-Mn, again resulting in perforation for some tubes.[21]

Sulfide-polluted seawater is increasingly rare because harbors are being cleaned up, but sulfide pollution can occur when systems are left full of stagnant seawater for extended periods of time (> 48 hours). When the system is started, there will be a mixture of deaerated seawater with sulfide and fresh aerated seawater, which is very aggressive. If there are dead legs that have been stagnant, the sulfide-polluted water can be slowly drawn into the aerated seawater stream over a period of days or weeks, creating extended aggressive conditions. The author has seen such attack on copper-nickel heat exchanger tubes in an MSF plant (see Chapter 6).

Ammonia can sometimes be present as a pollutant in seawater and it can have a number of sources. Ammonia may be present in the discharge water of fertilizer plants, it can come from the decomposition of organic material, and it can come from concrete with amine additives, or by the mass death of certain marine species. The concentrations can typically be from 0.5–10 mg/L, and ammonia has been observed to cause attack of heat exchanger tubes, particularly under deposits. The attack is characterized by metallic copper in the corrosion pits.[22] Heat transfer is necessary for pitting to occur, and 90/10 copper-nickel is the most susceptible to this kind of attack.[22] Aluminum brass with ferrous sulfate dosing is the copper alloy most resistant to ammonia attack. These concentrations of ammonia are insufficient to cause SCC of aluminum brass.[22] Accelerating the formation of protective films with ferrous sulfate dosing can prevent pitting due to ammonia, and some service failures in MSF plants and the methods to prevent attack are discussed in Chapter 6.

Ferrous sulfate dosing improves the corrosion resistance of aluminum-brass tubes to erosion corrosion and has prevented ammonia pitting of 90/10 copper-nickel. The optimum method for effective ferrous sulfate dosing is described in Appendix B.

Ammonia is well known to cause SCC of brasses under some circumstances. However, in seawater, extremely high concentrations of ammonia (≥ 1,000 mg/L) are required to cause SCC of aluminum brass, and the copper-nickel alloys are not susceptible at all.[13] However, high-strength copper-nickel (UNS 72400) has been observed to suffer SCC in a range of common chemicals found on offshore platforms and potentially in land-based plants.[13] Such chemicals include amine paints and lubricants, ammonia, natural seawater above 50 °C, and MoS_2-containing lubricants. For this reason, its use for marine fasteners has declined somewhat.

4.4 Stainless Steels

This discussion will be confined to the austenitic and duplex stainless steels that are routinely used in seawater in desalination plants. These are principally the 6% Mo austenitic and the superduplex alloys, but data is included for some of the lower alloyed stainless steels for comparison. Lower alloyed austenitic and duplex stainless steels are used in thermal desalination plants in areas with reduced dissolved oxygen, and in aerated seawater with cathodic protection (see Chapters 6 and 7).

The general corrosion rate for stainless steels in seawater is very low, typically 0.001 mm/y, but they are susceptible to various types of localized corrosion. Stainless steels do not normally pit in seawater, although welds can pit when used beyond the limits discussed below, or if poorly made (see Chapter 3).

Crevice corrosion is the main corrosion risk for stainless steels in seawater and it occurs when some of the metal is screened from the bulk seawater, restricting the ingress of fresh liquid. It occurs because of an oxygen concentration cell and the anode is inside the crevice. The deeper and tighter the crevice, the more likely it is that attack will occur. Some typical crevices in service are flanged joints, screwed couplings, and seals on shafts. The limits of use are affected by both temperature and the addition of chlorine/hypochlorite to control fouling.

The low-alloy stainless steels have crevice corrosion limits that are mainly a function of chloride concentration and temperature in aerated solutions. Figure 4.16 shows the limits from crevice corrosion tests on a range of austenitic and duplex stainless steels. Although not shown in Figure 4.16, the limits for 904L are similar to those for 2205. These limits increase in waters with low dissolved oxygen and these stainless steels find applications in thermal desalination plants, as described in Chapters 6 and 7.

Figure 4.17 shows the depth of crevice attack in natural seawater from several sources,[23-25] using crevice washers, as described in ASTM G78.[26] Ferralium is a 25% Cr duplex stainless steel with a PREN around 37 (UNS S32550). It can be seen that only the 6% Mo austenitic and superduplex alloy (Z100) resisted attack. Figure 4.18 shows the depth of crevice corrosion with ~ 1 mg/L chlorine at two temperatures.[24] The high-alloy stainless steels both resisted crevice corrosion at 16 °C, but only superduplex stainless steel (Z100) resisted attack at 40 °C. Francis and Byrne combined service data with laboratory test data to produce some recommended maximum operating temperatures at different chlorine concentrations for superduplex stainless steel, as shown in Table 4.8.[27] It can be seen that at low seawater temperatures the alloy can tolerate high chlorine concentrations. However, most desalination plants operate in warm seawater, and a chlorine concentration of 0.5 mg/L is usually more than adequate to control fouling. The data presented above show why NORSOK[28] (the Norwegian oil and gas standard) and some company standards insist on a PREN > 40 for stainless steels to be used in seawater.

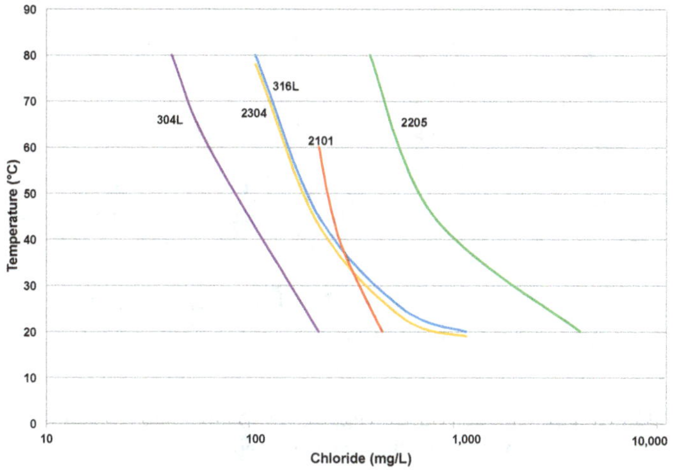

FIGURE 4.16 Crevice corrosion limits for some low-alloy stainless steels in aerated solutions.

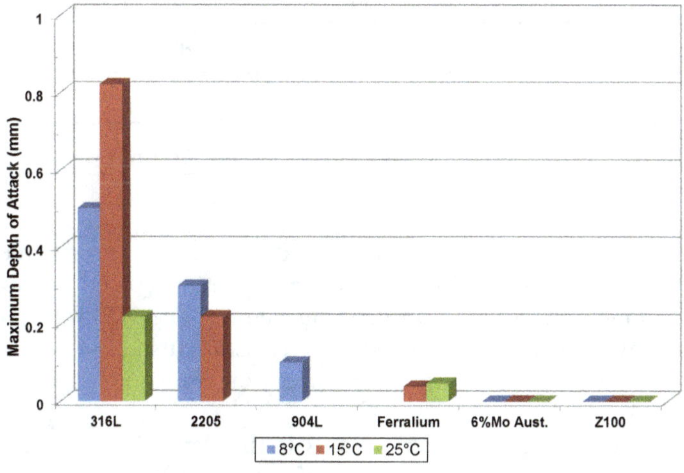

FIGURE 4.17 Maximum depth of crevice corrosion for some stainless steels at various temperatures.[23-25]

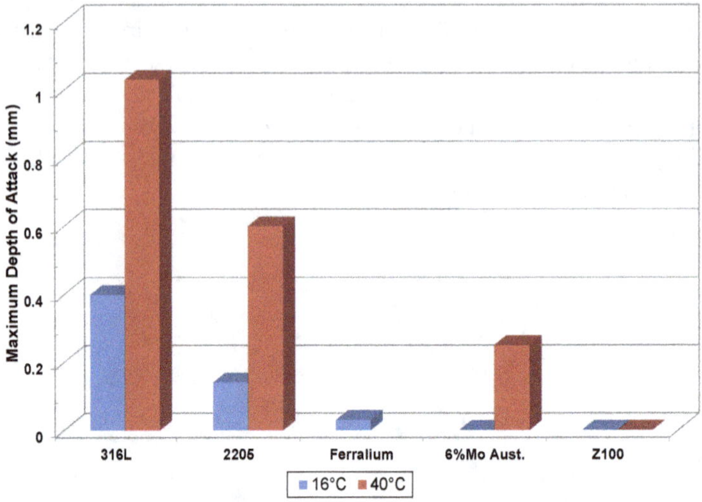

FIGURE 4.18 Maximum depth of crevice corrosion for some stainless steels in seawater with 1 mg/L chlorine.[24]

TABLE 4.8 Maximum operating temperatures for superduplex stainless steel as a function of chlorine content.[25]

CHLORINE (mg/L)	TEMPERATURE (°C)
200	10
5	20
1	30
0.5	40

The different minimum PREN values for different grades of stainless steel to resist crevice corrosion in warm seawater were discussed in Chapter 2. Ferritic stainless steels with a PREN > 35 (see Appendix A) will resist crevice corrosion in seawater up to about 35 °C, as discussed in Chapter 2, but these alloys are not usually used in desalination plants.

A common crevice in seawater systems is the flanged joint; Francis and Byrne reviewed a lot of test data with different gasket materials and developed some guidelines to avoid crevice corrosion.[29]

A. Low Pressure Systems (up to 10 bar).
   1. Avoid the use of polytetrafluoroethylene (PTFE) or graphite-loaded gaskets.
   2. Use gaskets made of synthetic rubber, rubber bonded aramid, or synthetic fiber.

B. High Pressure Systems (10–100 bar)
   1. Avoid PTFE coated gaskets.
   2. Graphite-containing gaskets are acceptable provided the graphite is sealed from the seawater and is never wetted.
   3. *Only* metals compatible with high-alloy stainless steel should be used for spiral wound gaskets. i.e., superduplex, 6% Mo austenitic, Ni-Cr-Mo alloys (where Mo > 7%), and titanium.

Seawater is filtered in seawater reverse osmosis (SWRO) plants to remove organic matter and this has an effect on the performance of stainless steels, particularly with regard to crevice corrosion. This is discussed in detail in Chapter 5.

In thermal desalination plants, some parts operate at low dissolved-oxygen contents, such as the heat recovery section in MSF plants. This means that lower alloy stainless steels can be used successfully; this is discussed with respect to MSF and MED plants in Chapters 6 and 7.

Stainless steels can be susceptible to SCC in seawater if the stress and temperature are sufficiently high. Figure 4.19 shows the threshold temperature for some stainless steels loaded to the 0.2% proof stress in aerated 5% sodium chloride.[30] The threshold temperature for chloride SCC as a function of chloride concentration is shown in Figure 4.20.[30] The threshold temperature for 304L is the same as that for 316L. It can be seen that SCC will not be a problem in SWRO plants, and only for 316L at temperatures ≥ 60 °C in thermal desalination plants. SCC of 304L and 316L can occur in thermal desalination plants if the temperature is above 60 °C, even when the environment inside is deaerated. If the external surface is wetted by condensation or splashing, then external SCC can occur. The solution is to use waterproof insulation coatings or replace the austenitic stainless steel with a duplex alloy, such as 2304 or 2205, both of which have superior resistance to SCC (Figure 4.20).

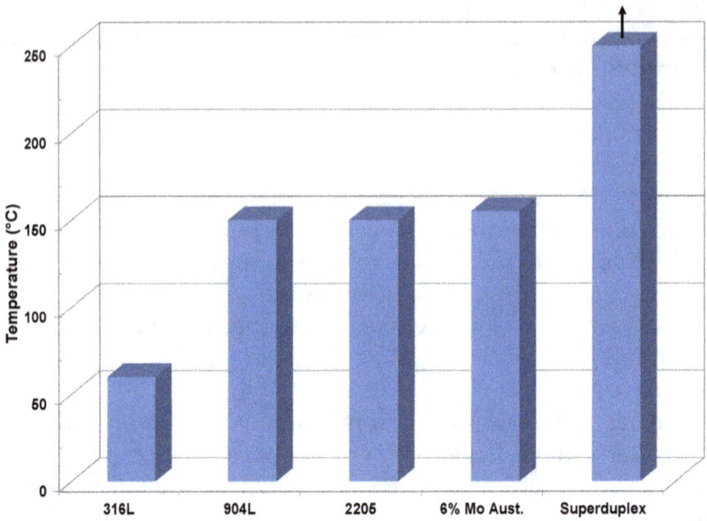

FIGURE 4.19 Threshold temperatures for SCC for some stainless steels in 5% NaCl loaded to the 0.2% proof stress.[30]

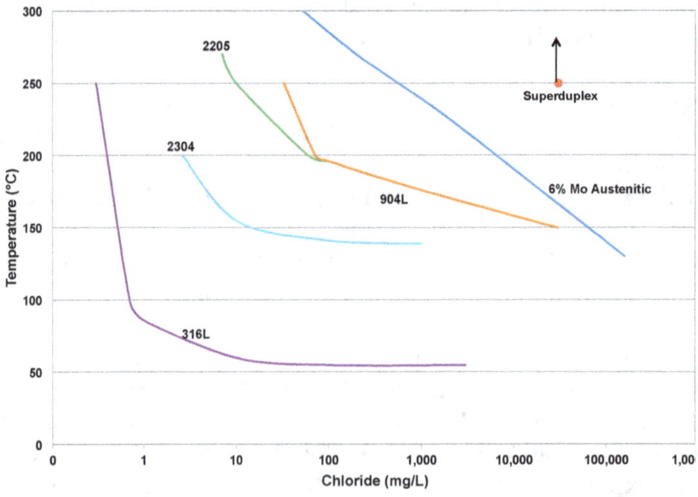

FIGURE 4.20 Threshold temperature for SCC of some stainless steels as a function of chloride concentration.[30]

Stainless steels have very good resistance to erosion corrosion in seawater. Kain et al. reported the depth of attack in seawater at 30 m/s and 30 °C for 30 days, as shown in Table 4.9.[31]

Table 4.9 Depth of attack in seawater at 30 m/s and 30 °C for 30 days according to Kain et al.[31]

Type of Alloy	Depth of Attack (mm)
316L	0.06
Superduplex	0.00
6% Mo Austenitic	0.00
M-Bronze	0.81

The stainless steels were all superior to M-bronze (see Appendix A) and, of the stainless steels, only 316L suffered any attack. Because of its good corrosion resistance combined with high strength, superduplex stainless steel has good resistance to erosion in the presence of silt and sand. Francis and Byrne describe how service data was used to infer practical limits of velocity and sand content for superduplex.[19] They also give some case histories, where the model was used to determine whether superduplex would be satisfactory, and service experience has shown the use of superduplex stainless steel to be justified.

Microbially induced corrosion (MIC) is well known to cause pitting of 300-series stainless steels and it has caused attack of alloys such as 2205 duplex and 904L austenitic in higher chloride waters.[30] Superduplex and 6% Mo stainless steels do not generally suffer from MIC, but Ohashi et al. describe tests on a range of stainless steels, where they were exposed to stagnant seawater rich in SRB at 30 °C for 30 days followed by exposure to clean seawater at the same temperature.[32] All of the stainless steels suffered corrosion, including crevice corrosion of superduplex. Parallel tests with pre-exposure in sterile seawater at 30 °C showed no attack of superduplex on subsequent exposure to flowing seawater. It was presumed that the sulfide from the SRB prevented the formation of stable protective films, such that crevice corrosion could initiate under flowing conditions. This suggests that even high-alloy stainless steels should not be left in stagnant biologically active seawater for extended periods prior to full start up. A case history where this is thought to have occurred in an SWRO plant is described in Chapter 5.

One area that requires care with lower alloy austenitic stainless steels is pipes and other equipment exposed to the atmosphere. Many desalination plants are situated in regions with high ambient temperatures, and by their nature they are close to the sea, so chlorides are present. Atmospheres at desalination plants are typically classified as "CX" (the most aggressive) as defined by ISO 9223.[7] Hot pipes in 316L that get condensation on them, or get splashed with seawater may suffer external chloride SCC. If insulation is used, it is important that it be one that does not absorb moisture, and that drainage holes are provided on the lower surfaces.

For equipment that is close to ambient temperature in the atmosphere, it is still possible for pitting and staining to occur on lower alloy stainless steels if they are close to the shore and exposed to high temperatures. References 33–36 describe the results of exposure tests of a range of stainless steels in both temperate and tropical climates for both urban and marine locations. These tests show that in the most demanding marine locations, 316L is not adequate and 2205 duplex stainless steel offers much higher corrosion resistance.

Where low-alloy stainless steels (such as 304L, 316L, and the lean duplex stainless steels) have been used for storage vessels, piping, etc., they may suffer staining, which can be removed by wire brushing, flapper wheels, or proprietary cleaners. However, if deep pitting occurs, it might be necessary to apply a coating to avoid further attack and possible leakage.

Addis et al. evaluated 10 coating systems for pipes and vessels made from 300-series stainless steels used offshore in the Gulf of Mexico, and found only 3 were adequate.[37] All three used a zinc-rich primer, two coats of epoxy paint, and a polyurethane topcoat for UV protection. They also found that the surface preparation was very important: the primer needed to be applied by hand brushing, rather than by spraying to ensure proper coverage in the pits.

4.5 Nickel Alloys

The nickel-copper alloys, such as alloys 400 and K-500, corrode similarly and the uniform corrosion rate in seawater is < 0.025 mm/y. The Ni-Cr-Mo alloys, such as alloy 625, corrode at a similar rate to stainless steels, i.e., ~ 0.001 mm/y.

Nickel-copper alloys have often been regarded as resistant to pitting in seawater, but this is probably because they are usually coupled to a less noble alloy that provides galvanic protection. Pitting of alloy 400 in natural seawater was observed in exposure tests at the LaQue laboratory,[38-39] in exposures of up to six months. Pitting was significantly less in seawater containing 1 mg/L chlorine. This was explained in terms of pitting potentials and open circuit potentials under both conditions.[38] Severe pitting of an alloy 400 seawater-piping system was observed by Shell in natural seawater.[40] It was also found that a continuous addition of ~ 0.8 mg/L chlorine reduced the open circuit potential below the pitting potential and prevented attack.[40]

The Ni-Cr-Mo alloys with PREN > 40 are very resistant to pitting in seawater, even in the presence of pollutants such as H_2S. However, the nickel-copper alloys will pit in the presence of sulfide pollution. The author has seen catastrophic pitting of a K-500 pump shaft in sulfide-polluted water that led to fatigue crack initiation at the base of the pits, where the stress concentration was highest.

The nickel-copper alloys are also susceptible to crevice corrosion in seawater, which becomes much deeper if it is coupled to a high-alloy stainless steel or a Ni-Cr-Mo alloy. The author bolted plates of alloy K-500 both to each other and to plates of Z100 superduplex stainless steel and observed crevice corrosion with all the test samples after 60 days exposure in natural seawater (Figure 4.21). Crevice corrosion can be prevented if the nickel-copper alloy is coupled to a sufficient area of a copper alloy or iron/steel.

FIGURE 4.21 The surfaces of a K-500/superduplex stainless steel couple, exposed to natural seawater, after separation (superduplex on left; K-500 on right). [Photo courtesy of Rolled Alloys]

All of the Ni-Cr-Mo alloys being discussed here have a PREN > 40, so they will resist crevice corrosion in warm seawater up to 40 °C, and higher for the higher alloyed materials. The exception to this is alloy 625, which has been observed to suffer crevice corrosion at 30–35 °C in seawater.[41] This is not often seen in service because the alloy is often connected to a sufficient area of an alloy that will provide cathodic protection, such as a copper alloy or uncoated carbon steel. The crevice corrosion resistance of nickel-alloy welds was discussed in Chapter 3.

The nickel-copper alloys have very good resistance to erosion corrosion in seawater, with a metal loss rate of < 0.025 mm/y at velocities from 1.5–40 m/s. The Ni-Cr-Mo alloys perform similarly to stainless steels, with no measurable metal loss up to 40 m/s.

Because of their high nickel content the nickel-copper and Ni-Cr-Mo alloys are not susceptible to chloride SCC in desalination environments.

The nickel-copper alloys are susceptible to pitting in deaerated seawater if the SRB become active,[13] but the Ni-Cr-Mo alloys are extremely resistant to MIC, particularly in the presence of SRB. Wade et al. reported pitting—which was very deep in crevices—of K-500 exposed in open seawater for 3 months.[42] The penetration rate was up to 3 mm/y for creviced samples, but was only up to 0.5 mm/y for non-creviced samples. They detected SRB and sulfides in the corrosion products and ascribed the attack to MIC.

4.6 Titanium

Titanium and its alloys are very resistant to all types of corrosion in seawater and the factor holding back their more widespread use is the high cost.[2] Titanium alloys can suffer crevice corrosion in chloride solutions under extreme conditions, as shown in Figure 4.22. It is clear that the commonly used grades, 2 and 5, will not suffer crevice corrosion at temperatures less than 70 °C in desalination environments.

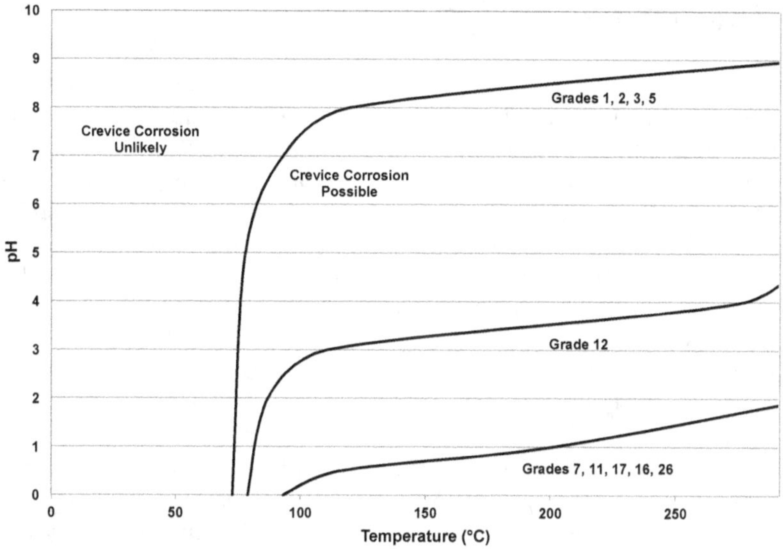

FIGURE 4.22 Effect of pH on threshold temperature for crevice corrosion of some titanium alloys.

Titanium is very susceptible to crevice corrosion in the presence of fluorides. Although fluorides are not usually found in desalination plants it is important to avoid fluorinated gasket materials (such as PTFE), because fluoride has leached from such gaskets and caused leakage of a titanium heat exchanger. There are alternative gasket materials that do not contain fluoride, such as synthetic rubber or synthetic fiber.

Despite its cost, titanium grade 2 has been used for heat exchanger tubes in thermal desalination plants when conditions are very corrosive for copper alloys, or high reliability is required. This use will increase when titanium prices are closer to those of copper alloys. It is important to realize that titanium has a lower fatigue resistance compared with copper alloys. This means that in a tube and shell heat exchanger, titanium tubes will require more baffle support plates than copper alloy tubes. This should not be a problem in a new heat exchanger, but fatigue failures have occurred where titanium tubes have been used to replace corroded copper alloy tubes, and adequate support for the tubes has not been provided.

Titanium is also much more electropositive compared with copper alloys, so it will cause galvanic corrosion of copper alloy tube sheets. This problem and the solutions are discussed in the next section.

4.7 Galvanic Corrosion

Galvanic corrosion occurs when two dissimilar metals are connected together in a conducting corrosive medium. The electropositive member of the couple stimulates attack on the electronegative member of the couple, due to the potential difference between the two alloys. The severity of the attack is controlled by the current density between the two metals and there are a number of factors that affect this, as discussed below. Galvanic corrosion and methods to avoid it are discussed in detail in Reference 43, and this should be consulted for further information over and above what is in this section.

Figure 4.23 shows the typical potentials for some common alloys in seawater at 10 °C, while Figure 4.24 shows the potentials at 40 °C. It is apparent that the potential difference can vary from ~ 20 mV to nearly 2 V. However, the magnitude of the potential difference alone will not indicate whether there will be galvanic corrosion or not. Corrosion is controlled by the current density (Faraday's Law), so anything that affects the current density will influence galvanic corrosion. The two main factors that do this are area ratio and cathodic efficiency. There have been instances of alloys with a potential difference of 800 mV being successfully coupled together and others of severe galvanic corrosion with a potential difference of only 50 mV.[43]

The anode is where metal dissolution is occurring, while the primary reaction on the cathode is the reduction of dissolved oxygen. The larger the anode to cathode area ratio, the smaller will be the current density on the anode. This is successfully employed, for example, in NAB valves with nickel-alloy trim. The large anode (valve body) to cathode (valve trim) area ratio means that the galvanic current density on the NAB is negligible from an engineering point of view.

If the metal forming the cathode is not very efficient at reducing dissolved oxygen, then the current density on the anode will not be too high, provided the anode is not smaller in area than the cathode. High-alloy stainless steels and Ni-Cr-Mo alloys form a biofilm in natural seawater in anything from 2 days to 2 weeks, depending on the local conditions. This biofilm is very efficient at reducing dissolved oxygen and means that lower alloy materials coupled to these alloys can suffer severe galvanic corrosion. However, when the seawater is dosed with chlorine/hypochlorite, to give a residual of ~ 0.5 mg/L, the high-alloy stainless steels and Ni-Cr-Mo alloys are much less efficient cathodes. An example of this is the coupling of 90/10 copper-nickel piping to high-alloy stainless steel. In natural seawater, the corrosion of the 90/10 copper-nickel is greatly accelerated at the junction of the two metals in natural seawater, but is hardly affected when the water

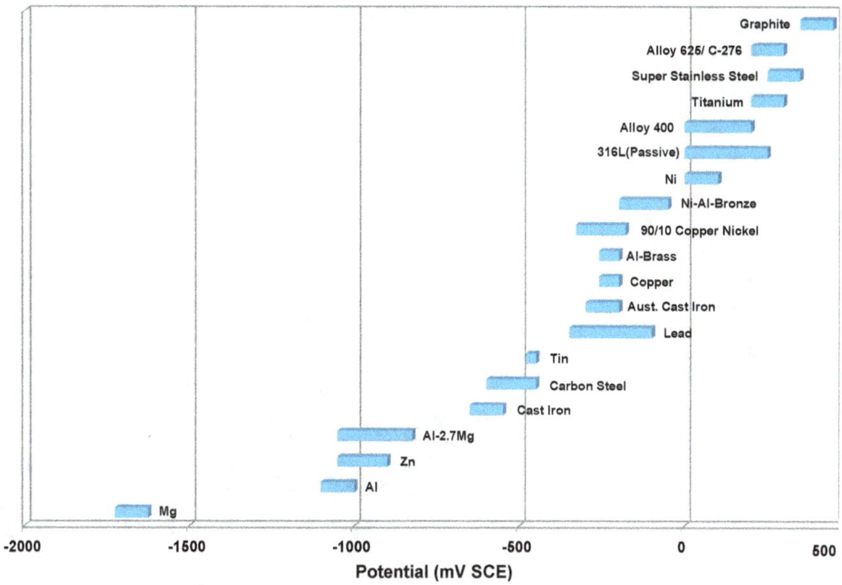

FIGURE 4.23 Galvanic series in seawater at 10 °C. [Data courtesy of SINTEF and the author]

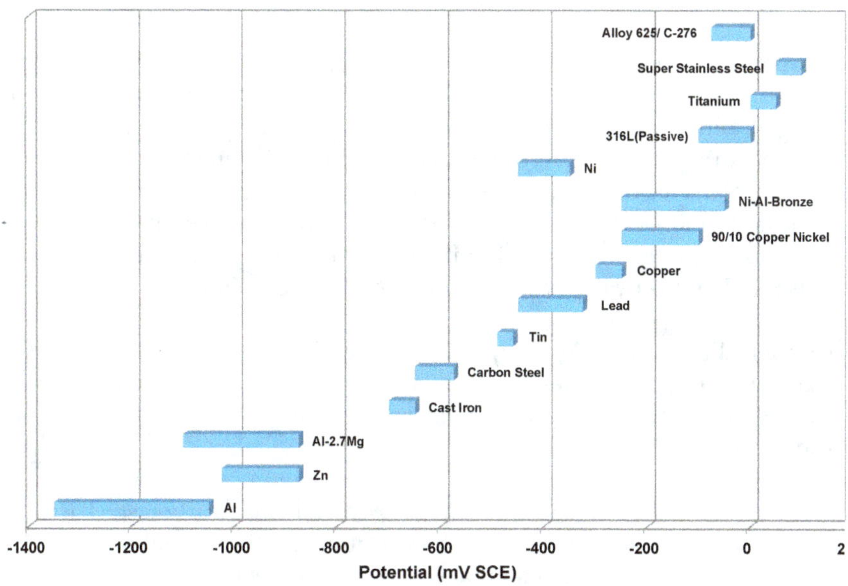

FIGURE 4.24 Galvanic series in seawater at 40 °C. [Data courtesy of SINTEF and the author]

is chlorinated, as shown in Figure 4.25. This is despite the potential difference being ~ 800 mV in chlorinated seawater.

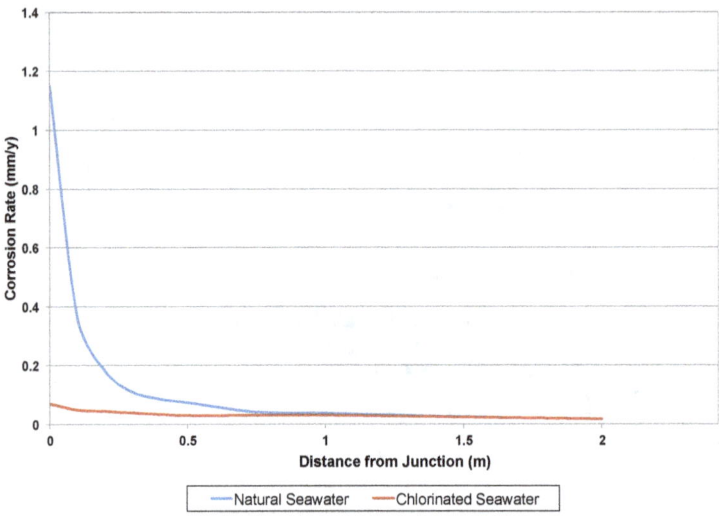

FIGURE 4.25 Corrosion rate versus distance from junction for 90/10 copper-nickel pipe coupled to a highly alloyed stainless steel pipe.[43]

Graphite is a very efficient cathode and it also has a very electropositive potential, as shown in Figure 4.23. If graphite loaded gaskets are used with flanged joints in a stainless steel chlorinated seawater system, galvanic corrosion can create severe crevice corrosion of the stainless steel.[29] This has led to leakage of both 6% Mo and superduplex flange joints. Graphite-free gaskets should be used, as described in Section 4.4 above. In spiral wound gaskets, graphite is acceptable provided it is never wetted. For the same reason, glands on valves and pumps should use graphite-free packing.

Copper alloys and nickel-copper alloys are also efficient cathodes in seawater and Francis points out that carbon steel coupled to copper-nickel suffers more attack than if coupled to the same area of titanium despite the greater potential difference with titanium.[43]

Figure 3.1 in Chapter 3 shows severe corrosion of a carbon-steel weld in poorly deaerated seawater due to the selection of a filler metal with a potential that is more negative than that of the parent metal. The potential difference was only ~ 50 mV, but the large cathode-to-anode area ratio produced a high current density on the weld metal, causing the severe metal loss.

One other factor that is important in thermal desalination plants is aeration, because the lower the dissolved oxygen content, the lower is the current from the cathode. In the parts of MSF and MED plants that operate at low oxygen contents (≤ 20 ppb) galvanic corrosion is not a major risk unless the anode-to-cathode area ratio is very small. However, where control of dissolved oxygen content is not good, and the concentration rises, not only does the corrosion rate of carbon steel increase, so does galvanic corrosion, such as at carbon steel/stainless steel junctions.

There is a wealth of data from galvanic corrosion studies in seawater (such as References 44–47) that there are groups of metals that can safely be connected together. These are shown in Table 4.10, where alloys in the same group can be safely coupled together. The exception is in the group 4 alloys, where cast iron and carbon steel are in a separate category to aluminum alloys; this is because it is known that couples between steel and aluminum alloys will cause galvanic corrosion of the aluminum alloy. This is for ambient temperature seawater where, for instance, all the alloys in group 1 are passive. This means that high-alloy stainless steel piping can be connected to titanium heat exchangers, provided there are no alloys in the heat exchanger from another group.

TABLE 4.10 Alloy groupings for galvanic compatibility in seawater at ambient temperature.

CATEGORY	TYPE	ALLOY
1	Noble; passive	Nickel-Chrome-Moly Alloys (Mo > 7%) 6% Mo Austenitic Stainless Steel Superduplex Stainless Steel Titanium and its alloys
2	Passive; not truly corrosion resistant	Alloy 400/ K-500 904L 22% Cr Duplex Alloy 825 Alloy 20 316L
3	Moderate Corrosion Resistance	Copper Alloys Austenitic Cast Iron
4	Poor Corrosion Resistance	Carbon Steel Cast Iron
		Aluminum Alloys

There are also examples where mixing alloys from different groups can cause accelerated attack of the anodic metal. A good example is coupling NAB valves to superduplex stainless steel piping, where the author showed that severe corrosion of NAB would occur in natural seawater and at a 10:1 cathode-to-anode area ratio in chlorinated seawater.[48] The service experience is that severe corrosion of NAB valves in superduplex piping systems occurred on both the Woodside Goodwyn A and BP Andrew platforms. The valves are usually thick wall castings so the corrosion manifests as the valve being unable to seal properly rather than through-wall penetration. The seawater on Goodwyn A was chlorinated, but the seawater temperature varied from 20–40 °C, i.e., higher than in the author's tests.[48] The seawater on BP Andrew was unchlorinated at about 10 °C, and severe corrosion occurred within one year. The main valves were replaced in superduplex stainless steel and the others were changed as they failed. Chlorination was also retrofitted because stainless steel is a less efficient cathode with chlorine, as discussed above, and this slows down the rate of attack.

The Eastern Trough Area Project (ETAP).in the North Sea also used superduplex stainless steel seawater piping and NAB valves, but with chlorination from the outset. The valves also corroded, but more slowly than on Andrew, and the valves were changed to superduplex stainless steel after about five years. Figure 4.26 shows a corroded NAB butterfly valve from a chlorinated superduplex stainless steel seawater piping system. Some operators have made NAB valves in superduplex stainless steel piping systems work by using internally coated stainless steel spool pieces on either side of each valve, as discussed below. However, the spool pieces need to be 5 or 6 pipe diameters long, so there must be space to incorporate them.

FIGURE 4.26 Corroded NAB butterfly valve from a superduplex seawater piping system.

There are a number of ways to prevent or minimize galvanic corrosion. The main ones for seawater are discussed below, but there is more detail in Reference 43. For a new build, it is best to select alloys all from the same group to avoid problems. With desalination plants, this will usually either be group 1 or group 3 alloys. It is possible to use alloys from different groups in some instances, but there must be data to support their successful use when they are used together. With retrofits it can be more complicated, but selecting alloys from the same group generally avoids galvanic corrosion problems.

With piping systems, one method to prevent galvanic corrosion is the insulating gasket at a flanged joint. However, there are two potential problems that can make it ineffective. The first is that the insulation joint must be correctly assembled to ensure electrical isolation of the two pipes and this requires skilled labor. The second is that the pipes are often connected externally, either through the electrical earthing system or via pipe hangers/supports, both of which will prevent the isolation flange from working. It is important to check for electrical continuity between the two pipes before the system is filled with water, to ensure the isolation flange will work.

An alternative is to use an isolation spool pipe, which may be made of a polymer, such as FRP, or it can be a metal pipe of the cathode alloy with a high quality coating on the inside and up the flange faces. It must not be made of the anode-piping alloy because there will be greatly accelerated attack at any small defects in the coating. For copper alloys and stainless steel, the spool piece is usually 5–6 pipe diameters long. When isolating stainless steel from carbon steel, the spool piece is typically 2.0–2.5 m long; hence, there must be adequate space to fit these spool pieces and adequate space for access to fit them.

In MSF flash chambers, there is sometimes a need to coat sensitive areas with stainless steel and it is important to avoid galvanic corrosion of carbon steel at the junction. One method is to coat the stainless steel up to about 25–50 mm from the junction, using a high quality coating, such as synthetic rubber, which will give an adequate life without introducing extra maintenance.

Another method of preventing galvanic corrosion is to use cathodic protection, usually as sacrificial anodes, because impressed current systems need to be very robust to work reliably for many years in desalination plants. An example is the increasing use of low-alloy stainless steels for tube sheets with titanium tubes. Alloys such as 316L, 2101, and 2304 will suffer accelerated corrosion in seawater when coupled to titanium, as shown in

Figure 4.27. Figure 4.27A shows the voluminous corrosion products between the tubes, while Figure 4.27B shows the deep pitting between the tubes after cleaning. This is normally prevented by inserting sacrificial low-carbon steel anodes in each water box (as discussed in Chapter 7). In this case, however, the anodes had accidentally been electrically isolated from the tube sheet and water box. It is important to use a low-carbon (< 0.15% C) steel so that fragmentation does not occur and the anodes corrode more or less uniformly. It is also important to size the anodes so that they will last between maintenance intervals for the plant, without causing undue turbulence in the water box. It is common to size anodes to last for two years, when there will be less than 50% of the thickness left.

(A)

(B)

FIGURE 4.27 Appearance of 316L tube sheet with titanium tubes in the final heat exchanger in an MED plant: A) as opened, B) after cleaning, showing deep attack.

Where titanium or high-alloy stainless steel tubes are retrofitted to a heat exchanger with copper-alloy tube plates, there will be severe attack of the tube sheet unless it is cathodically protected, as shown in Figure 4.28. As for low-alloy stainless steels, the tube sheets can be protected with low-carbon steel anodes, with the same provisos. The use of zinc or aluminum alloy anodes should be avoided, because these will over-protect the tube sheets and can cause scale build up just inside the entrance to the tubes.[43]

FIGURE 4.28 Galvanic corrosion of NAB tube sheet with titanium tubes.

4.8 Polymers

Polymers were discussed in Chapter 2 and it was stressed that with FRP it is important to select a veil that will resist corrosion in the environment. In seawater, epoxy resins are usually adequate, but if there is silt or sand in the cooling water, it is important to specify an erosion resistant veil as well. Under very high silt or sand burdens, it may be necessary to switch to superduplex stainless steel or titanium.

Polymers such as unplasticized polyvinyl chloride (uPVC) and chlorinated PVC (CPVC) are sometimes used for piping to handle chemicals, usually being dosed into some part of the seawater/brine system. It is important to ensure that the PVC will resist the chemical being transported at the local ambient temperature, which may be high, such as in the Middle East. Manufacturers usually have data sheets showing what chemicals can be handled safely and the temperature limit.

REFERENCES

1. Melchers, R.E., "Effect of Temperature on the Marine Immersion Corrosion of Carbon Steels," *Corrosion* 58, 9 (2002): p. 768.
2. Powell, C., and R. Francis (eds.), "The Corrosion Performance of Metals for the Marine Environment: A Basic Guide," *European Federation of Corrosion Publication No. 63* (London, United Kingdom: Maney Publishing, 2012).
3. Berger, F.P., and K.-F.F.-L. Hau, "Mass Transfer in Turbulent Pipe Flow Measured by the Electrochemical Method," *International Journal of Heat and Mass Transfer* 177, 20, 11 (1977): pp. 1185–1194.
4. Farinha, P.A., *Sub-Sediment Corrosion of Steel Sheet Piling in Ports and Harbours with Particular Reference to Sulphate Reducing Bacteria*, Ph.D. Thesis (Manchester, United Kingdom: University of Manchester, 1982).
5. Francis, R., G. Byrne, and H.S. Campbell. "The Corrosion of Some Stainless Steels in a Marine Mud," CORROSION 1999, paper no. 313 (Houston, TX: NACE International, 1999),
6. Volkening, V.B., "Corrosion of Steel Pipe by Chlorinated Seawater at Different Velocities," *Corrosion* 6, 4 (1950): p. 123.
7. ISO 9223 (latest revision) "Corrosion of Metals and Alloys-Corrosivity of Atmospheres-Classification, Determination and Estimation" (Geneva, Switzerland: International Standards Organization).
8. International Nickel, Ltd.,"Ni-Resist Austenitic Cast Irons; Properties and Applications" (New York, NY: International Nickel Ltd., 1965).
9. CDA, "Aluminum Bronze Alloys Corrosion Resistance Guide," publication 80 (Potters Bar, United Kingdom: Copper Development Association, 1981).
10. ASTM G32 (latest revision), "Standard Test Method for Cavitation Erosion Using Vibratory Apparatus" (West Conshohocken, PA: ASTM International).
11. Miyasaka, M., and N. Ogure, "Stress Corrosion Cracking of Austenitic Cast Irons in Seawater and Brine, and Its Prevention," CORROSION 1986, paper no. 324 (Houston, TX: NACE International, 1986).
12. Francis, R., *The Selection of Materials for Seawater Cooling Systems: A Practical Guide for Engineers* (Houston, TX: NACE International, 2006).
13. Francis, R., *The Corrosion of Copper and Its Alloys: A Practical Guide for Engineers* (Houston, TX: NACE International, 2010).
14. Schreiber, C.F., O. Osborn, and F.H. Coley, "Corrosion of Metals in Desalination Environments," *Materials Protection* 7, 10 (1968): p. 20.
15. Oldfield, J.W., G. Swales, and B. Todd, "Corrosion of Metals in Deaerated Seawater," 2nd ed., Middle East Corrosion Conference, January 1981 (Houston, TX: NACE International; and Manama, Bahrain: Joint Bahrain Society of Engineers, 1981): p. 48.

16. Ross, R.W., "The Effect of Seawater Chemistry on the Corrosion of Cu-Ni-Fe Alloys at Elevated Temperatures," CORROSION 1977, paper no. 94 (Houston, TX: NACE International, 1977).
17. Bog, S., "The Resistance to Impingement Attack by Seawater of Copper Alloy Tubes," 5th Scandinavian Corrosion Congress, Copenhagen, Denmark, 1968.
18. Bog, S., "The Resistance of Copper Alloys to Different types of Corrosion in Seawater," 7th Scandinavian Corrosion Congress, Trondheim, Norway, 1975.
19. Francis, R., and G. Byrne, "The Erosion Corrosion Limits of Duplex Stainless Steels," *Materials Performance*, 57, 5 (2018): p. 44.
20. Francis, R., "The Effect of Additions to the Cooling Water on the Corrosion of Copper Alloy Condenser Tubes," *Materials Performance*, 21, 8 (1982): p. 44.
21. Francis, R., "Effect of Pollutants on Corrosion of Copper Alloys in Seawater. Part 2: Sulfide and Chlorine," *British Corrosion Journal* 20 (1985): p. 175.
22. Francis, R., "Effect of Pollutants on Corrosion of Copper Alloys in Seawater. Part 1: Ammonia and Chlorine," *British Corrosion Journal* 20 (1985): p. 167.
23. Shone, E.B., R.E. Malpas, and P. Gallagher, "Stainless Steels as Replacement Materials for Copper Alloys in Seawater Handling Systems," *Transactions of the Institute of Marine Engineers* 100 (1988): p. 193.
24. Francis, R., "Effects of Chlorine on Corrosion of High Alloy Stainless Steel in Seawater," Stainless Steel '87, September 1987 (London, United Kingdom: Institute of Metals): p. 192.
25. Kain, R.M., "Crevice Corrosion Resistance of Duplex Stainless Steels in Chloride-Containing Waters," Duplex Stainless Steels '97 (Zutphen, Netherlands: KCI Publishing, 1997): p. 627.
26. ASTM G78 (latest revision), "Standard Guide for Crevice Corrosion Testing of Iron-Base and Nickel-Base Stainless Steels in Seawater and Other Chloride-Containing Environments" (West Conshohocken, PA: ASTM International).
27. Francis, R., and G. Byrne, "Experiences with Superduplex Stainless Steel in Seawater," *Stainless Steel World*, 16 (Zutphen, Netherlands: KCI Publishing, 2004): p. 53.
28. NORSOK M-001, "Materials Selection, Edition 5" (Oslo, Norway: Standards Norway, 2014).
29. Francis, R., and G. Byrne, "Gasket Selection for Stainless Steels in Seawater," *Materials Performance* 46, 10 (2007): p. 50.
30. Francis, R., *The Corrosion of Duplex Stainless Steels: A Practical Guide for Engineers* (Houston, TX: NACE International, 2018).

31. Kain, R.M., L.S. Marshall, D.G. Melton, D.M. Aylor, R.A. Hays, and R.J. Ferrara, "Corrosion, Galling, and Wear Testing of Duplex and Austenitic Stainless Steels for Seawater Valve Applications," *Stainless Steel World* (December 2000): p. 28.
32. Ohashi, K., R. Kobayashi, J.F.D. Stot, and M.J. Schofield, "Marine Crevice Corrosion of Stainless Steels Under Biofilmed and Sterile Conditions," CORROSION 2016, paper no. 7109 (Houston, TX: NACE International, 2016).
33. Lojewski, C., P. Boillot, and J. Peultier, "Atmospheric Corrosion Resistance of Duplex Stainless Steels: Results of a Field Exposure Program," Duplex 2010, Session IV.B.5 (Zutphen, Netherlands: KCI Publishing, 2010).
34. Le Manchet, S., and D. Paul, "Atmospheric Corrosion Resistance of Stainless Steels for Architecture," CORROSION 2017, paper no. 8880 (Houston, TX: NACE International, 2017).
35. Mameng, S.H., R. Pettersson, L. Wegrelius, and C. Leygraf, "Effect of Surface Finish on the Atmospheric Corrosion of Duplex Grade UNS S32205; Results of a Field Exposure Program in Dubai," CORROSION 2016, paper no. 7231 (Houston, TX: NACE International, 2016).
36. Mameng, S.H., and L. Wegrelius, "Atmospheric Corrosion Resistance of Stainless Steel in the Middle East: Result of a Field Exposure Programme in Various Environments," CORROSION 2017, paper no. 8964 (Houston, TX: NACE International, 2017).
37. Addis, T.T., A. Bodington, C. Fonseca, and B.T.A. Chang, "Evaluation of Offshore Coating System on Pitted Stainless Steel, CORROSION 2018, paper no. 100487 (Houston, TX: NACE International, 2018).
38. Tipton, D.G., and R.M. Kain, "Effect of Temperature on the Resistance to Pitting of MONEL Alloy 400 in Seawater," CORROSION 1980, paper no. 36 (Houston, TX: NACE International, 1980).
39. Klein, P.A., R.A. Hayes, R.J. Ferrara, and R.M. Kain, "Effect of Electrolytic Chlorination on the Crevice Corrosion of 70/30 Copper-Nickel and Nickel-Copper Alloy 400," CORROSION 1991, paper no. 509 (Houston, TX: NACE International, 1991).
40. Gallagher, P., A. Nieuwhof, and R.J.M. Tausk, "Experiences with Sea Water Chlorination on Copper Alloys and Stainless Steels," in *Marine Corrosion of Stainless Steels–Chlorination and Microbial Effects*, EFC Publication No. 10 (London, United Kingdom: The Institute of Materials, 1993): p. 73.
41. Martin, F.J., P.M. Natishan, K.E. Lucas, E.A. Hogan, A.M. Grolleau, and E.D. Thomas, "The Influence of Environmental Factors on the Crevice Corrosion of Alloy 625 in Natural Seawater," CORROSION 2002, paper no. 213 (Houston, TX: NACE International, 2002).

42. Wade, S., A. Ang, R. Piola, and W. Niel, "Microbially Influenced Corrosion of UNS N05500," *Materials Performance* 57, 4 (2018): p. 66.
43. Francis, R., *Galvanic Corrosion: A Practical Guide for Engineers*, 2nd ed. (Houston, TX: NACE International, 2017).
44. Walleen, B., and T. Andersson, "Galvanic Corrosion of Copper Alloys in Contact with a Highly Alloyed Stainless Steel in Seawater," *Proceedings of 10th Scandinavian Corrosion Congress* (Stockholm, Sweden: Swedish Corrosion Institute, 1986): p. 149.
45. Francis, R., "The Galvanic Corrosion of Zeron 100 and Alloy K-500 in Seawater," Technical Report TN1099 (Temperance, MI: Rolled Alloys, 1994).
46. Shone, E.B., and P. Gallagher, "Galvanic Compatibility of Selected Alloy Stainless Steels in Seawater," in *Corrosion in Seawater Systems*, A.D. Mercer (ed.) (London, United Kingdom: Ellis Horwood, 1990): p. 40.
47. Griffiths, A., and A. Turnbull, "Galvanic Coupling of Corrosion Resistant Alloys in Seawater at Elevated Temperatures," *Stainless Steel World* (Jan/Feb 1998): p. 38.
48. Francis, R., "Coupling of Superduplex Stainless Steel and Cast Nickel-Aluminum Bronze in Seawater," *British Corrosion Journal* 34, 2 (1999): p. 139.

5 Reverse Osmosis Desalination

The principle of reverse osmosis is simple. If low-chloride and high-chloride water are separated by a semipermeable membrane, water will transfer through the membrane from the low chloride side to the high chloride side. This process is called osmosis. The transfer of water in one direction then builds up a net pressure on the high chloride side; this is called the "osmotic pressure." If a pressure is applied to the high chloride side that exceeds the osmotic pressure, then water will transfer faster from the high chloride to the low chloride side. This process is called "reverse osmosis" and the transferred water is relatively pure compared to the original brine. Commercial units use a whole series of membranes, which consist of either a sheet-like membrane wound around an inner core, or numerous hollow, fine fibers laid in parallel that selectively block the transfer of chloride ions. The membrane cells are used in series and parallel to produce the required water quality. Reverse osmosis is most commonly used to remove the chloride from seawater (seawater reverse osmosis or SWRO), but it may also be used to purify brackish waters of lower or higher chloride content. One advantage of SWRO is that it can cover a wide range of water requirements, from a small unit to supply a hotel, to a large multitrain unit to supply a city, as shown in Figure 5.1.

5.1 Method of Operation

All SWRO plants follow the same type of layout, as shown schematically in Figure 5.2. They can be divided into four sections.

(A)

(B)

FIGURE 5.1 Typical SWRO desalination plants: A) small unit for a hotel; B) large multitrain unit for a city. [Photos courtesy of Rolled Alloys]

The first section is the low-pressure seawater intake, where the maximum pressure is 10 bar and the seawater is filtered and treated. This is necessary because the membrane cells are the single highest cost item in a reverse osmosis plant and it is important to prevent degradation of the membranes to extend their life. The seawater may have chlorine or hypochlorite injected near the seawater intake to prevent fouling. It then passes through coarse filters to remove plastic bags, fish, and other large items. The seawater is then usually dosed with a chemical to aid coagulation as it passes through the dual media filters. These are large vessels with layers of various materials to remove organic matter and other material from the seawater. Finally, the water is fine filtered through cartridge filters to ~ 1 μm. The effects of finer filtering are discussed in Section 5.6.

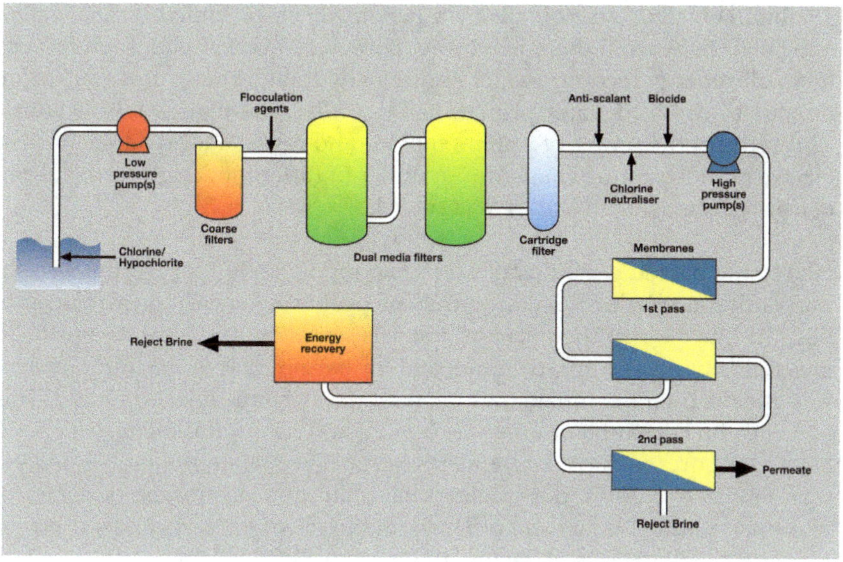

FIGURE 5.2 Schematic diagram of a typical SWRO plant.

Depending on the pH of the feed water, it would be normal to add an anti-scalant into the suction line of the high pressure pumps to limit precipitation of calcium and magnesium salts. Chlorine and hypochlorite are powerful oxidizing agents and they will cause damage to the membranes, unless they are removed. Removing these agents is usually done by monitoring the redox potential, and dosing with sodium metabisulfite to bring the redox potential down to an acceptable value, as discussed in the next section. It is also common to add a nonoxidizing biocide to limit organic fouling of the membranes.

The second, or high pressure, section can operate from 60–100 bar, with 70 bar being a typical value, while high recovery systems may operate closer to 100 bar. In the membrane cells chloride is removed in two or more passes, depending on how pure the final water must be.

After chloride removal, the third section handles the permeate (pure water), and is around 1 bar pressure; it is then pumped to storage or a distribution system. Although this is low pressure and low chloride, it is quite corrosive because the water has little tendency to form protective scales of calcium carbonate.

The fourth section is the reject brine section, where the chloride concentration may be twice that of natural seawater. The pressure is still high (~ 69 bar) and this energy is not wasted, but is recovered prior to discharge.

Traditionally, this was done using a Pelton wheel, or similar device, generating electricity from the reject brine. However, modern plants tend to use hydraulic energy recovery, which transfers hydraulic energy from the brine directly to the feed water prior to, or just after, the high pressure pump. This reduces the power of pump required and is much more efficient than electricity generation. There are a number of different designs using different materials, as discussed in Section 5.4.3.

Finally, the oil and gas industry uses a variant on SWRO for treating certain waters that are being reinjected into oil-producing strata to increase production rates. When the injection water is seawater, it is deaerated prior to injection, but if there are strontium and/or barium salts in the water present with the oil, then the sulfate must be removed from the injection water. This prevents the formation of insoluble barium and strontium sulfate, which will block the pores in the oil producing strata. The sulfate removal process is very similar to SWRO, but it uses a nanofiltration membrane designed to selectively remove sulfate and other di- and trivalent ions. Materials selection is constrained by the same factors that govern SWRO plants.

5.2 Conditions in SWRO Plants

5.2.1 Low Pressure Section

This is virtually identical to any system handling low pressure seawater, and materials must be selected as if for standard seawater service. Because many SWRO plants operate in warmer climates, the seawater inlet temperature will often exceed 20 °C, and may be as high as 40 °C in areas such as the Arabian Gulf in summer.

Copper alloys are not often used in this section, because excessive copper ion concentrations may affect membrane performance and may exceed WHO recommendations for copper in drinking water. However, nickel aluminum bronze (NAB) has been used for seawater lift pumps in some plants.

Much more common is the use of stainless steel and it is important to remember that only high-alloy stainless steels will resist crevice corrosion under these conditions. Standards such as NORSOK M-001,[1] the Norwegian oil and gas standard, say that stainless steels and nickel alloys for seawater service must have a pitting resistance equivalent number (PREN) > 40.0, where

$$\text{PREN} = \%\text{Cr} + 3.3(\%\text{Mo} + 0.5\times\%\text{W}) + 16\times\%\text{N}$$

In waters with chloride contents lower than seawater ($\leq$ 4,000 mg/L), then lower alloy stainless steels may be acceptable, as discussed in Chapter 4.

Carbon steel may be used with a suitable long-life coating for some applications, but the quality of the coating is essential to guarantee a reasonable service life. Such coatings include synthetic rubbers, such as neoprene, and glass flake epoxy.

High chromium and molybdenum nickel alloys (such as alloys 625 and C-276) and titanium and its alloys, are rarely used because of their high cost. Fiberglass reinforced plastic (FRP) is used in larger plants for low pressure seawater piping, particularly in larger diameters and where there are long runs with few bends and tees.

5.2.2 High Pressure Section

All the filtration and treatment of the water for the high pressure section makes it less corrosive than natural seawater. The membranes will degrade if the redox potential exceeds 350 mV Ag/$AgCl_{SAT}$, but dosing with sodium metabisulfite to achieve very low redox potentials is expensive so the practical lower limit is 250 mV Ag/$AgCl_{SAT}$. In most SWRO plants, the redox potential is close to the upper limit. Byrne et al. showed that the range of 250–350 mV Ag/$AgCl_{SAT}$ corresponds to an open circuit potential for stainless steels of 100–200 mV SCE (saturated calomel electrode).[2] In practice, stainless steels in most plants will have a potential close to the upper limit.

Figure 5.3 shows the wide range of potentials that are possible for stainless steels in seawater.[2] In natural seawater, a biofilm forms over a period of a few days to a few weeks, which depolarizes the cathodic reaction (the reduction of dissolved oxygen), elevating the potential to around +300 mV SCE. If chlorine or hypochlorite is added to the seawater to control fouling, the biofilm does not form, but chlorine and hypochlorite are powerful oxidizers and the potential rises to around +600 mV SCE. If the seawater is heated to 25–30 °C above the normal ambient temperature, the biofilm does not form and the potential is then much lower, around +150 mV SCE. If the seawater is partially deaerated, then the potential falls further, and with 200 ppb oxygen it is around –100 mV SCE. When the seawater is fully deaerated the potential is around –450 mV SCE. The red lines in Figure 5.3 show the maximum and minimum potentials for stainless steels in the high pressure section of an SWRO plant.[2]

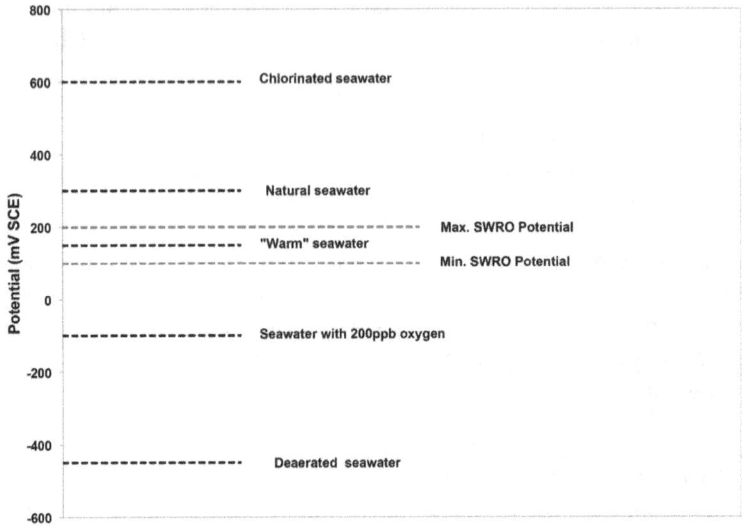

FIGURE 5.3 Possible potentials for stainless steels in seawater.[2]

Byrne et al. conducted crevice corrosion tests on a range of stainless steels over the full range of potentials shown in Figure 5.3.[2] Their results, in Figure 5.4, show that alloy 316L is totally unsuitable in seawater at any potential down to −100 mV SCE. Alloys 904L and 2205 were only resistant up to around 22 °C at potentials of 200 mV SCE or greater, but there was a sharp increase in the critical crevice temperature (CCT) at lower potentials. Most applications of 904L and 2205 in the high pressure section of SWRO plants have failed because of crevice corrosion and/or pitting of welds,[2–3] because most plants operate at higher temperatures than 22 °C. Where these alloys have worked, the temperatures have been lower, or the open circuit potential was closer to the lower limit in Figure 5.3. Hence, despite all the treatment of the seawater, only superduplex and 6% Mo austenitic alloys will perform satisfactorily under the conditions commonly found in the high pressure section.

Byrne et al. also investigated the effect of chloride on the CCT at different potentials for 2205 and Z100 superduplex stainless steel, as shown in Figure 5.5. It is clear that the CCT is hardly affected at all by chloride, with a reduction of only 10–20 °C over the range 3,000–100,000 mg/L chloride. It can also be seen that lower potentials produce a much greater increase in CCT for superduplex compared with 2205. It is expected that the results for the 6% Mo alloys should be similar to superduplex, while those for 904L should be similar to 2205.

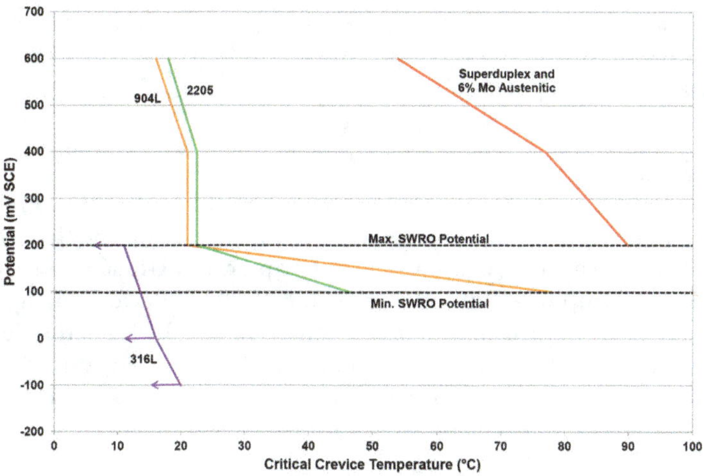

FIGURE 5.4 Critical crevice temperatures for some stainless steels as a function of potential.[2]

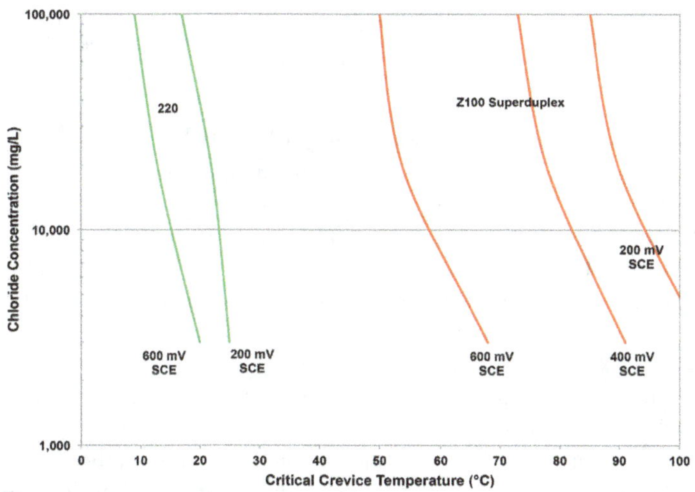

FIGURE 5.5 Critical crevice temperatures for some stainless steels as a function of chloride.[2]

5.2.3 Reject Brine

The chloride concentration in the reject brine section may be double that in the seawater feed, but this will not affect the choice of materials, because superduplex and 6% Mo austenitic stainless steels will be more than adequate, while lower alloy stainless steels will not, as discussed in the previous section (see Figure 5.5). This assumes there have been no changes in the water treatment prior to the high pressure pumps that have significantly changed the redox potential.

5.2.4 Permeate

The selection of materials for this section will depend on the final chloride level of the permeate. As discussed in Chapter 4, 304L stainless steel will be satisfactory up to ~ 100 mg/L chloride at 20 °C. Because many SWRO plants are in warmer climates, it is common to use 316L for the permeate piping, because of its greater resistance to chlorides in warmer waters. Polymer pipes and fittings may also be used, but if the water is for drinking, the polymer must be approved for potable water.

5.3 Components in the Low Pressure Section

5.3.1 Pumps

Seawater typically enters larger plants down a large diameter pipe or a channel to a seawater basin and there are usually strainers at some point before the basin. The pipes may be FRP or lined concrete, while open channels are usually lined concrete. The low pressure seawater pumps sit in the basin, where they feed water to the rest of the plant. These pumps usually operate most of the time, but one pump may be on standby in larger units.

In older plants, NAB has occasionally been used for these pumps, but accelerated corrosion can occur with NAB if it spends some time in stagnant seawater and sulfate reducing bacteria (SRB) become active (see Chapter 4). If NAB pumps are used, it is imperative that they are not connected directly to high-alloy stainless steel piping, because this will cause accelerated corrosion of the NAB due to galvanic corrosion,[4] as discussed in Chapter 4.

It is well established that cast stainless steels such as 2205 duplex (ASTM A995, grade 4A) suffer crevice corrosion in warm seawater, similar to the wrought alloy (see Chapter 4). It is also well established that 25% Cr duplex stainless steels with a PREN from 34–37 can also suffer crevice corrosion in SWRO plants.[5] Hence, it is common for the seawater lift pumps to be in cast superduplex stainless steel to ASTM A995, grade 5A or 6A. Such pumps have generally given good service. Because of its composition, grade 5A tends to be more ferritic and therefore more prone to cracking in thicker sections. Francis and Hebdon discuss the merits of castings in both grades and conclude that larger castings are preferred in grade 6A.[6] However, the quality of the castings is most important and this is discussed in more detail in Chapter 8. Francis and Hebdon also discuss castings in the 6% Mo austenitic grades, and point out the problems with obtaining sound castings.[6] In addition, the lower strength and higher molybdenum and nickel contents of these alloys make castings more expensive than those in one of the superduplex grades.

In larger SWRO plants, the seawater lift pumps were sometimes made of austenitic cast iron with stainless steel impellers and shafts. However, there have been numerous failures of such pumps by stress corrosion cracking (SCC) in warmer waters, as described in Chapter 4. Many of these have been replaced with superduplex stainless steel pumps, and the use of austenitic cast iron for seawater pumps has declined.

5.3.2 Valves

In large diameter piping in concrete or FRP, it is common to use butterfly valves that are rubber lined steel with an NAB disc. These will work fine provided that the valve is not in direct contact with a high-alloy stainless steel pipe.[4] It is also important that the valve does not spend excessive times in the closed position while full of seawater. If SRB become active, the sulfide produced can cause rapid pitting of the NAB (see Chapter 4).

Smaller valves and all those fitted in high-alloy stainless steel piping should be in a high-alloy stainless steel to avoid galvanic corrosion problems, and superduplex is usually the most cost effective choice. As for pumps, it is important that the castings be of good quality if the valve is to perform properly without corroding (see Chapter 8).

One thing that is sometimes neglected are minor valves, such as those used in drains. They are often positioned at the bottom, with a short pipe between the main system and the valve. The stagnant conditions adjacent to the valve can result in the production of sulfides by SRB, and this makes an aggressive solution when mixed with the aerated water above it. Larche et al. report the failure of a NAB disc in a coated steel drain valve due to sulfide pitting.[5]

The solution is to use high-alloy stainless steel (PREN > 40) for all minor valves in contact with seawater to avoid failures.

5.3.3 Strainers and Filters

Inlet strainers and filters can be subjected to high local stresses when there are excessive solid bodies in the incoming water. For this reason, copper alloys are not widely used for inlet strainers. If 316L stainless steel is used, it must be recognized that it will suffer crevice corrosion in seawater and so it must be protected by cathodic protection. This may take the form of unpainted cast iron or steel housings, or sacrificial anodes. It is difficult to make a large cathodically protected filter system that is robust enough to survive without regular maintenance. Sometimes, 2205 duplex is used for strainers, and while it is not immune to crevice corrosion in seawater, it will last longer than 316L. An alternative is to use superduplex stainless steel, which is readily available as wires and mesh. In addition, thin sheets can be drilled or punched to make a filter with a specific pass size. The entire support structure can be superduplex, or polymers, where a lower strength is acceptable.

The dual media filter vessels can be large (several meters in diameter and height) and it is common to fabricate them from carbon steel and coat all internal surfaces with a synthetic rubber, such as neoprene. The rubber lining has a finite life, usually 10–15 years, after which relining is required. To avoid relining, it is necessary to fabricate the vessels from a stainless steel with good resistance to crevice corrosion in seawater, such as superduplex or 6% Mo austenitic. An alternative would be to line carbon steel filter vessels with 2 or 3 mm thick sheeting in a 6% Mo alloy, which is welded into place with a nickel-alloy filler (see Chapter 3). Both of these will increase the initial cost, but will avoid relining.

The final filters are usually polymeric and the bodies may be polymer as well, or carbon steel with a good quality coating, such as neoprene. For smaller filters, coating carbon steel is difficult, and in such cases these can be superduplex or 6% Mo austenitic stainless steel to avoid crevice corrosion problems.

5.3.4 Chemical Injection

Figure 5.2 shows several points where chemicals are injected prior to the high pressure pumps. The storage tanks for these chemicals may be polymer or stainless steel, but they must resist corrosion by the stored chemical. The advice of the chemical supplier is not always reliable. The author was involved with chemical storage tanks in 316L that leaked—despite 316L being recommended. The chemical was an acidic biocide (pH ~ 2.3); attack occurred at the waterline if the storage tank was not in use, that is, if the chemical level was not constantly moving. Corrosion tests showed that 316L would resist corrosion in the bulk chemical, but was susceptible to attack at the waterline at both 5% and 20% concentration, at 30 °C, a typical temperature in the Middle East. The tests also showed that 2205 duplex was resistant to waterline attack at 5% but not at 20%, while Z100 superduplex was resistant at both concentrations.

If FRP is used for storage tanks, it is essential that a resin and veil are chosen that will resist the specific chemical being stored. For example, glass is unsuitable with alkaline chemicals. This is discussed in more detail in Chapter 2.

The piping from the storage tank to the valve prior to the dosing point must resist the chemical, and is typically 316L stainless steel or a polymer, such as chlorinated PVC (CPVC) or unplasticized PVC (uPVC). The isolation valve immediately prior to the injection point must resist both the chemical and seawater, so it is usually a polymer or superduplex stainless steel. The pipe connecting the valve to the main line is usually stainless steel so that it is easily welded to the main seawater line. To resist corrosion it should be in superduplex stainless steel. Where a screwed coupling is used, then a polymer pipe connection is also possible.

5.3.5 Instrument Tubing

In all plants there is a requirement to monitor the conditions at various points of the process, and in an SWRO plant, all the water is aerated and flowing. The selection of an instrument tubing alloy is a perennial problem and one for which no clear guidelines exist. The problem is what tubing materials should be used with different main piping materials. Because small-bore tubing is only available in a limited number of materials, the selection to choose from is also limited

The recommendations presented here have been compiled from extensive discussions with offshore operators, corrosion consultants, and the author's extensive experience with galvanic corrosion. The four tubing materials that are principally considered are 316L stainless steel, 90/10 copper-nickel, superduplex, and 6% Mo austenitic stainless steels. Other materials, such as alloy 400 have been used, but 90/10 copper-nickel will work where alloy 400 works and it is less expensive and is widely available.

Because all the main water systems are flowing and aerated, there are just two possible conditions to consider. One is when the liquid to the sensor is static, such as for pressure gauges, and the other is when the water flows through the sensor, such as for chlorine monitors.

When the water to the sensor is static, it quickly becomes deaerated because the tube diameter is small, typically 10 or 12 mm (⅜ or ½ in.). Hence, as long as the stand-up height to the isolation valve is sufficient (see Figure 5.6), the tubing alloy after the valve does not need to be totally resistant to seawater. Because the main seawater pipe will be FRP or a high-alloy stainless steel (PREN > 40), there are only a few limited choices, as shown in Table 5.1. Alloy 316L is not resistant to aerated seawater, so the minimum stand-up height is important to ensure aerated water is not in contact with it. The other alloys are seawater resistant and a minimum stand-up height is not required.

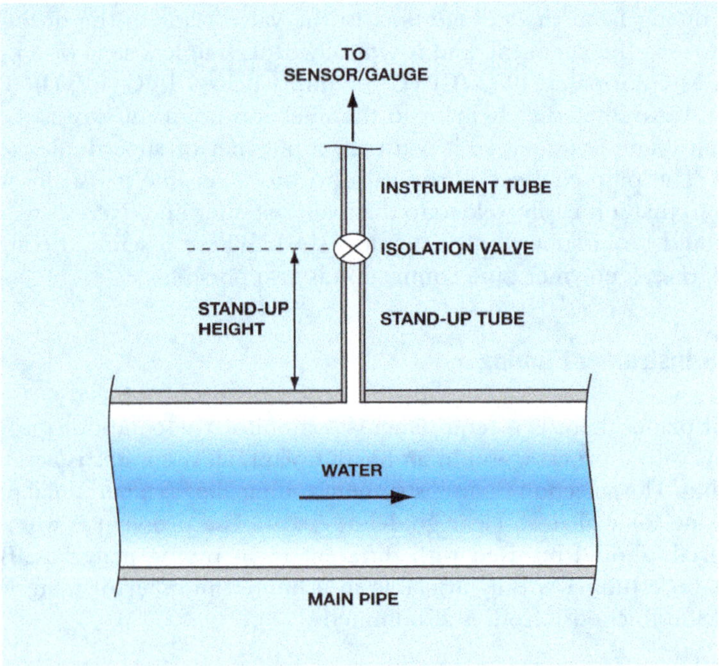

FIGURE 5.6 Schematic diagram showing takeoff for instrumentation tubing.

Where the seawater to the sensor is flowing, then the instrument tubing must be resistant to aerated seawater, and must also be galvanically compatible with the isolation valve alloy, as shown in Table 5.1. Alloy 316L stainless steel is not acceptable, but 90/10 copper-nickel is OK if it is not in contact with a high-alloy stainless steel, which will cause galvanic corrosion. There is no minimum stand-up height for flowing water in the instrument tubing.

TABLE 5.1 Recommended materials for instrumentation tubing in static and flowing, aerated seawater systems.

CONDITION IN SENSOR	MAIN PIPING ALLOY	STAND-UP PIPING	ISOLATION VALVE MATERIAL	MINIMUM STAND-UP HEIGHT (mm)	SENSOR TUBING ALLOY
STATIC	FRP	FRP/ Polymer	NAB	None	90/10 Cu-Ni *
				150	316L
	6% Mo/ Superduplex	6% Mo/ Superduplex	6% Mo/ Superduplex	150	316L
				None	6% Mo/ Superduplex
FLOWING	FRP	FRP/ Polymer	NAB	None	90/10 Cu-Ni *
			6% Mo/ Superduplex	None	6% Mo/ Superduplex
	6% Mo/ Superduplex	6% Mo/ Superduplex	6% Mo/ Superduplex	None	6% Mo/ Superduplex

* Preferred if external corrosion is a concern (see text).

Where both 90/10 copper-nickel and 316L stainless steel are acceptable, the copper-nickel alloy is preferred in cases where the 316L may suffer crevice corrosion under clamps, etc., that might be splashed with seawater, or where condensation might collect.

5.4 Components in the High Pressure Section

5.4.1 Pumps

Pumps in the high pressure section are multistage to deliver the high pressure required (60–100 bar) and are either a horizontally split case or a radially split case design for ease of maintenance. These are usually made of superduplex stainless steel because of its high strength, ease of manufacture, and lower cost compared with 6% Mo austenitic alloys. This alloy selection also avoids problems due to crevice corrosion, as discussed previously. As for the low pressure pumps, it is important that the castings are of good quality and the guidelines of Francis and Hebdon should be followed (see also Chapter 8).[6]

5.4.2 Piping

Piping has to resist pressures up to 100 bar and must also resist crevice corrosion in seawater. It is common to use superduplex stainless steel piping, and attempts to use lower alloy stainless steels have often resulted in severe pitting corrosion problems at welds and/or crevice corrosion problems at high pressure couplings. The 6% Mo austenitic grades are also acceptable, but are less used because of their higher cost. Because of the need to be able to remove any membrane cell for maintenance or replacement, pipe connections are made with high pressure couplings, as shown in Figure 5.7. Each pipe stub has a square cut groove about 15 mm from the end and a synthetic rubber boot (or seal) joins the two pipes. Two half clamps fit in the grooves and are bolted together to make a seal. These clamps are usually galvanized steel or 316L stainless steel, but a failure is described in Section 5.5 where fusion-bond, epoxy-coated steel clamps were used.

FIGURE 5.7 A typical high pressure coupling with one half-clamp removed for clarity.

The boot creates a tight crevice, and lower alloy stainless steels (PREN < 40) have suffered severe crevice corrosion in SWRO plants, as shown in Figure 5.8. Superduplex stainless steel would be expected to resist crevice corrosion under the boot of high pressure couplings, but Byrne et al. report corrosion under the rubber boot of Z100 superduplex in one plant.[2] They carried out crevice corrosion tests on the outside of samples of Z100 superduplex pipe with different surface finishes at 200 mV SCE–the maximum stainless steel potential in the high pressure section. The results in Figure 5.9 show that the maximum CCT was unchanged with surface finish, but the tube with the outer surface machined showed a much smaller scatter. This is thought to be because machining removed any ovality and roughness, both of which could create local crevices that were much tighter. Machining the land under the boot prevented corrosion there in subsequent SWRO applications.[2]

FIGURE 5.8 Crevice corrosion of 316L under the boot of a high pressure coupling.

The strength of superduplex stainless steel is such that it can easily retain the pressure as schedule 10S pipe in smaller sizes (up to the nominal pipe size [NPS] 8 at 70 bar and up to NPS 4 at 100 bar), rather than the normal schedule 40S.[7] However, there is then insufficient wall thickness to machine in the groove for the high pressure couplings.[7] It is possible to cold roll a groove into schedule 10S superduplex pipe to accommodate the coupling clamps, and this has been successfully applied to piping in two SWRO plants in the Philippines at 65 bar.[7] If the tool forming the rolled groove becomes worn, then the coupling may not seal properly, but some companies are developing a modified rolled grove that is more secure. As pressures in the high pressure section increase, it may not be possible to retain the pressure with sufficient safety margin with rolled grooves in schedule 10S, so schedule 40S pipe would then be essential.

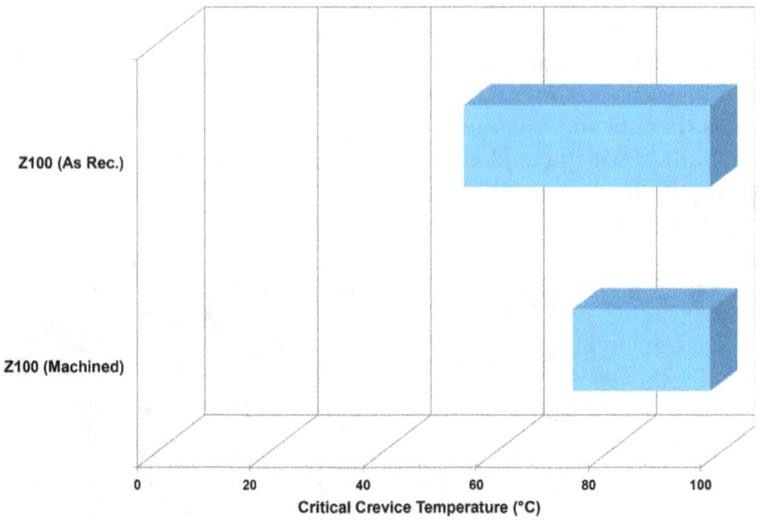

FIGURE 5.9 The effect of surface finish on the CCT range of superduplex in seawater at a potential of +200 mV SCE.[2]

Piping in the second pass may be in a lower alloy stainless steel if the chloride concentration is low enough (≤ 4,000 mg/L), and 2205 has been used successfully in larger plants. However, the amount of second pass piping in smaller plants is low and it is often more convenient to make it all in superduplex stainless steel, especially because it removes the risk of the wrong grade being used in the first pass.

In SWRO piping it is frequently required to take a small branch from a large diameter pipe, for example NPS 1.5 from an NPS 8 main pipe. One way is to first drill a small hole in the large pipe. Then, a ball is placed inside the pipe and is pulled through the small hole to create a stand up onto which another pipe can be welded. (This technique is known as "pulling a branch"). Superduplex stainless steel is difficult to pull in schedule 40S pipe and tearing may occur. An alternative is to drill a hole in the large pipe and fillet weld the branch to it. This is difficult to do, even in the fabrication shop, and corrosion has been seen in such fillet welds. Such a procedure really needs a corrosion test as part of the qualification to ensure the joint will be sound, as discussed in Chapter 8. The preferred method is to use a reduced branch tee fitting, which will only require regular butt welds to fit into the piping.

It is also common to use threaded connections for things such as drain plugs and sensors. Such joints create tight crevices and have been known to suffer crevice corrosion. It is important to use a screwed connection in superduplex stainless steel and seal the threads with both polytetrafluoroethylene (PTFE) tape and a suitable sealant, such as Loctite[1,2] brand sealants, to ensure water does not enter the threads.

The comments made in Section 5.3.5 about instrument tubing alloy selection for the low pressure section also apply to the high pressure section.

5.4.3 Membrane Pressure Vessels

The membrane pressure vessels are constructed in FRP and are suitably rated for operation up to 83 bar. The vessels are designed and manufactured as per ASME Section X.[8] The high pressure feed and reject ports are mostly in superduplex stainless steel. The industry standard is to use side-ported vessels with multiple ports. This allows economy in piping. The ports are grooved to allow use of flexible couplings for connection with high pressure piping. The industry is moving towards use of duplex and superduplex stainless steel couplings with U.S. Food and Drug Administration (FDA)-approved rubber gaskets.

5.4.4 Reject Brine Section

The reject brine section is at high pressure and has a high chloride content, but Figure 5.5 shows that this will not have a significant reduction in the CCT for superduplex and 6% Mo austenitic stainless steels. These alloys are generally used for the piping, but the hydraulic energy recovery devices may be made with ceramic internals, polymers, or superduplex stainless steel, depending on the design. The materials must resist corrosion, erosion, and maintain efficiency. Figure 5.10 shows a typical layout of superduplex reject brine piping coupled to hydraulic energy recovery devices. Couplings in this section are also of the high pressure type shown in Figure 5.7.

1. The listing of trademarks, brand names, manufacturers, etc., is for informational purposes only and does not constitute endorsement of these products or manufacturers.
2. Loctite is a registered trademark of Henkel Corporation, Rocky Hill, CT.

FIGURE 5.10 Superduplex brine reject piping coupled to hydraulic energy recovery devices. [Photo courtesy of Rolled Alloys]

5.4.5 Permeate Section

The selection of material for the permeate piping, valves, etc., will depend on the final chloride content, but it is common to use 304L or 316L stainless steel. In warmer climates and/or higher chloride concentrations, 316L is preferred. It is also possible to use plastic pipes and fittings, but these must be approved for drinking water if the permeate is potable water, and FDA-approved plastic pipes and fittings are becoming the norm for this service.

5.5 Service Experience

One SWRO plant in the Middle East wanted to avoid relining the dual media filter vessels after 10–15 years for a plant designed to last 25 years with a seawater temperature of ~ 30 °C. The plant was large and they fabricated several large vessels from 2205 duplex stainless steel. After 6 weeks they began to leak because of pitting at welds (Figure 5.11) and there was also crevice corrosion at flanged joints. This was only to be expected because 2205 only has a PREN of around 35 and is not resistant to crevice corrosion in seawater, as discussed above.

FIGURE 5.11 Leaks due to pitting at weld of 2205 duplex dual media filter vessel.

Some low pressure superduplex stainless steel pumps in another SWRO plant in the Middle East showed crevice corrosion after entering service, with a seawater temperature of 25–30 °C (Figure 5.12). The pumps had been mostly stagnant for the 6–9 months of final construction and commissioning, and were only run intermittently with very infrequent chlorine dosing. In addition, the pumps were left full of seawater when idle. Larche et al. attributed the corrosion to SRB activity during the extended stagnant period that resulted in less protective films, so crevice corrosion could initiate during service.[5] This was confirmed by Ohashi et al., who carried out crevice corrosion tests on a range of austenitic and duplex stainless steels in seawater.[9] Samples were exposed for three weeks at 30 °C in airtight jars containing a mixed bacterial culture containing SRB from the Arabian Gulf, so they would be used to warm environments. The samples were then exposed to natural aerated seawater at 30 °C for six months. Parallel tests were conducted on samples exposed only to sterile seawater. The results in Table 5.2 show that the prior exposure in a water-supporting biofilm development can greatly increase the susceptibility to crevice corrosion and pitting for lower alloyed materials. Even superduplex stainless steel suffered attack in these tests, while it is normally accepted that it is suitable up to around 40 °C in natural seawater (Chapter 4). The results also show the importance of the biofilm in assisting in crevice corrosion initiation, in that 2205 suffered no attack in the sterile seawater. These tests might be said to represent a worst case where a system is left full of stagnant seawater for several weeks with lots of nutrients for bacterial growth prior to operation in aerated warm seawater. This is exactly the scenario reported by Larche et al.;[5] it demonstrates the need to flush the system with fresh water and/or drain the pumps and piping prior to extended shutdown (> 48 hours).

FIGURE 5.12 Crevice corrosion of superduplex LP pump case after extended stagnation prior to service.

In another SWRO plant, there was severe corrosion of the low pressure superduplex pump casing and shaft (Figure 5.13). A detailed metallurgical investigation revealed the following problems:

- Parts of the pump did not meet the PREN > 40 requirement.
- There was sigma/chi phase in some places, due to poor heat treatment, where corrosion had occurred.
- The structure was very ferritic.
- There were major weld repairs that had not had any postweld heat treatment, contrary to ASTM A995.[10]
- The shaft also contained sigma/chi phase due to poor heat treatment, and had also corroded.

TABLE 5.2 Results of crevice corrosion tests on some stainless steels after prior exposure to seawater with a bacterial culture.[9]

ALLOY	BIOFILM ENVIRONMENT	MAX. DEPTH (mm)	WT. LOSS (mg)	STERILE ENVIRONMENT	MAX. DEPTH (mm)	WT. LOSS (mg)
304L	Very severe and extensive attack	1.15	469	Limited crevice corrosion	0.065(C)	2
316L	Very severe and extensive attack	1.29	218	Minor crevice corrosion, some pitting	0.059(P) 0.012(C)	2
2101	Very severe and extensive attack	>2	497	Very minor pits	0.021(P)	2
2205	Severe crevice corrosion	0.54	15	No Corrosion		1
2507	Some crevice corrosion	0.26	17	No Corrosion		1

(C) = Crevice Corrosion
(P) = Pitting Corrosion

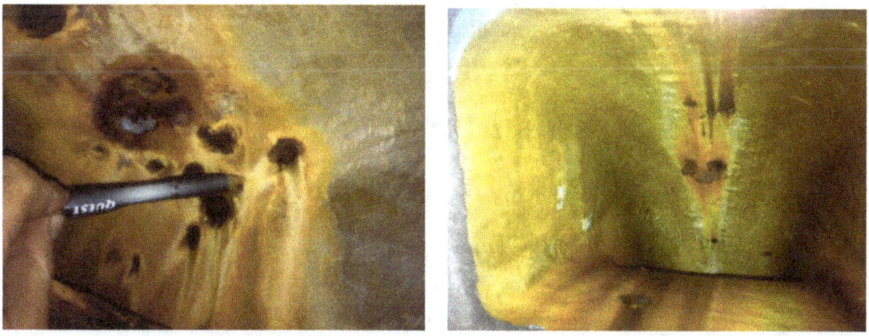

(A) (B)

FIGURE 5.13 Severe corrosion of superduplex stainless steel castings due to poor quality control: (A) pitting at major weld repair, and (B) pitting at sigma/chi phase due to poor heat treatment

The pump castings had been purchased to ASTM A995[10] grade 5A with no additional testing requirements and there was no quality assurance or quality control (QA/QC) monitoring or third party witnessing throughout production. The need for these to obtain good quality duplex and superduplex castings is discussed in more detail in Chapter 8.

Duplex stainless steels also require care in welding, and guidelines on joint design, heat input, and interpass temperature are available from suppliers of welding consumables (for example Ref. 11). In addition, there is good general advice on welding duplex stainless steels in an International Molybdenum Association publication that is available free from their website.[12] Poor quality welding often leads to the formation of sigma phase in the heat-affected zone (HAZ), and this reduces the corrosion resistance, leading to leaks, as shown in Figure 5.14. There is more information on welding in Chapter 3.

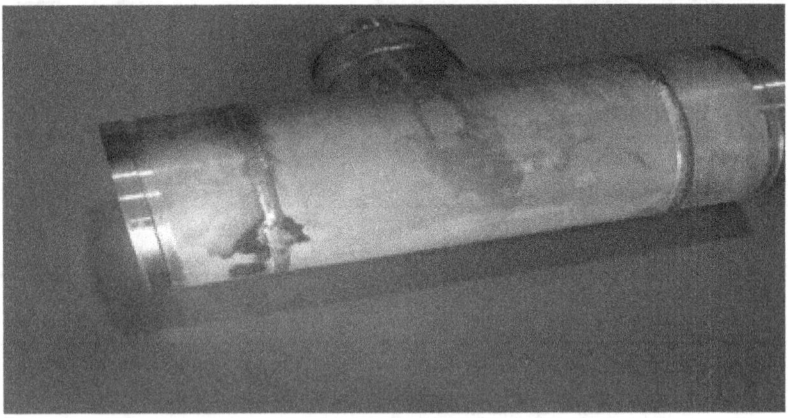

FIGURE 5.14 Leak at a poorly welded superduplex pipe joint. [Photo courtesy of Rolled Alloys]

The importance of selecting galvanically compatible valves was stressed above, and Figure 5.15 shows severe galvanic corrosion of an NAB drain valve fitted in a superduplex stainless steel high pressure line.

The necessity for avoiding the use of graphite-loaded gaskets was highlighted in Chapter 4, but in SWRO plants spiral wound gaskets are sometimes used, especially in the high pressure section. The metal for the windings must be compatible with the flange alloy, which usually means, superduplex stainless steel, 6% Mo stainless steel, Ni-Cr-Mo alloys with Mo > 7%, or titanium.[13] Sometimes a 316L spiral-wound gasket has been used, but coated with PTFE to prevent corrosion of the windings. PTFE forms a much tighter crevice than a metal spiral-wound gasket, and it can cause crevice corrosion when it would not be expected, as shown in Figure 5.16.[13] The conclusion is that PTFE-coated gaskets should be avoided.[13]

FIGURE 5.15 Galvanic corrosion of an NAB drain valve in a superduplex stainless steel HP line. [Photo courtesy of Rolled Alloys]

FIGURE 5.16 Crevice corrosion of superduplex stainless steel flange under a PTFE-coated spiral wound gasket. [Photo courtesy of Rolled Alloys]

An unusual problem occurred in the high pressure section of an SWRO containerized plant in the Middle East, where the high pressure couplings were fusion bond epoxy (FBE)-coated carbon steel. Corrosion was seen on the end of the feed pipes to the membrane cells and on the exit pipes from the cells.[14] This was stray current corrosion, which was thought to be due to the plant's location on land that has a high conductivity. The land was recovered from the sea (and hence had a high chloride content); and all the local electrical systems were earthed into this land. It was demonstrated that the FBE coating acted as an insulator, which meant that much of the metal piping was floating with respect to earth. The problem was solved by anchoring all the piping to a common earth.[14] This demonstrates that galvanized steel or stainless steel clamps are the best choice for the high pressure couplings, because these will ensure a continuous electrical connection between all the piping.

5.6 Future Trends

In current SWRO plants the final filters in the low pressure section generally go down to ~ 1 µm, but there are advantages in finer filtration, if it can be made reliable and economic. Microfiltration would be down to ~ 0.1 µm, while ultrafiltration would be to ~ 0.01 µm, and nanofiltration would be to ~ 0.001 µm. While micro or ultrafiltration could extend membrane life, they do not remove the need to inject antiscaling chemicals and biocides. The advantage of nanofiltration would be that chemical dosing of antiscalants and biocides prior to the high pressure pumps would be unnecessary and the finer filtration should also prolong membrane life. It is not known if this very fine filtration would further reduce the redox potential to enable lower alloyed stainless steels to be used. The author thinks it unlikely that the redox potential would be sufficiently reduced, and crevice corrosion testing would be advisable before going ahead with a low-alloy in the high pressure section, in a plant with nanofiltration.

REFERENCES

1. NORSOK M-001, "Materials Selection, Edition 5" (Oslo, Norway: Standards Norway, 2014).
2. Byrne, G., R. Francis, G. Warburton, and J. Wilson, "Electrochemical Potential and the Corrosion Resistance of Stainless Steels in SWRO Applications," IDA International Desalination Conference (Topsfield, MA: International Desalination Association, 2009).
3. Olsson J.O., and M.M. Snis, "Don't Repeat Mistakes! An SWRO Plant Case Study," IDA International Desalination World Congress, paper no. SP050-036 (Topsfield, MA: International Desalination Association, 2005).
4. Francis, R., *Galvanic Corrosion: A Practical Guide for Engineers*, 2nd ed. (Houston, TX: NACE International, 2017).
5. Larche, N., P. Dezerville, and D. Le Flour, "Corrosion and Corrosion Management Investigations in Seawater Reverse Osmosis Desalination Plants," *Desalination and Water Treatment* 51 (2013): p. 1744.
6. Francis, R., and S. Hebdon, "The Selection of Stainless Steels for Seawater Pumps," CORROSION 2015, paper no. 5446 (Houston, TX: NACE International, 2015).
7. Byrne, G., R. Francis, G. Warburton, R.J. Bullock, and C. Kuzler, "The Selection, Design, Fabrication and Performance of Zeron 100® in SWRO Applications," IDA International Desalination Conference (Topsfield, MA: International Desalination Association, 2003).
8. ASME (latest revision), *Boiler Pressure Vessel Code, Section X, Fiber Reinforced Pressure Vessels* (New York, NY: ASME International).
9. Ohashi, K., R. Kobayashi, J.F.D. Stot, and M.J. Schofield, "Marine Crevice Corrosion of Stainless Steels Under Biofilmed and Sterile Conditions," CORROSION 2016, paper no. 7109 (Houston, TX: NACE International, 2016).
10. ASTM A995, *Standard Specification for Castings, Austenitic-Ferritic (Duplex) Stainless Steel for Pressure-Containing Parts* (West Conshohocken, PA: ASTM International).
11. Rolled Alloys, "Welding Guidelines for ZERON 100," Publication No. 105, 2009, https://www.rolledalloys.com/shared-content/technical-resources/fabrication-information/ZERON-100_FI_US_EN.pdf (December 2018).
12. IMOA, *Practical Guidelines for the Fabrication of Duplex Stainless Steels*, 3rd ed. (London, United Kingdom: International Molybdenum Association, 2014).
13. Francis, R., and G. Byrne, "Gasket Selection for Stainless Steels in Seawater," *Materials Performance* 46, 10 (2007): p. 50.
14. Byrne, G., R. Francis, and A. Gulamhusein, "Stray Current Corrosion In Sea Water Reverse Osmosis Systems," *Proc. International Desalination Association World Congress* (Topsfield, MA: International Desalination Association, 2015).

6 Multistage Flash (MSF) Distillation

Multistage flash desalination (MSF) is a simple process where fresh water is condensed from vapor released from seawater. As described in Chapter 1, the process has been used for many years.

6.1 Method of Operation

The MSF process is basically similar to distillation by boiling. Water can be made to boil just as effectively by reducing the pressure as by raising the temperature. In a closed vessel, the temperature and pressure are roughly proportional so that a decrease in pressure can cause instantaneous boiling of some of the water, with the characteristic "flashing" off of some of the water vapor.

An MSF plant consists of a series of chambers (often 20 or more), each operating at a pressure lower than the preceding one. As heated brine flows from one chamber to the next, some of it flashes off into vapor at progressively lower pressures. The vapor passes through demisters to remove any entrained brine droplets and then condenses on cooler condenser tubes. The distillate then drops into collection trays and passes from stage to stage in a distillate channel, prior to distribution as drinking water. This usually involves the injection of CO_2 (sometimes in the form of incondensable gases extracted from the distiller) to lower the pH. This distillate is then passed through limestone beds to increase hardness and finally has a small quantity of chlorinated seawater added to provide some minerals and also to disinfect the final product.

The basic MSF process is shown schematically in Figure 6.1. Steam heats the recirculating brine in the brine heater and the steam condensate produced is returned as feed to the boilers. The hot brine (typically 110 °C) then passes through a series of flash chambers in the heat recovery section, each at a lower pressure than the preceding one. In each chamber, some water vapor flashes off, as described above. Figure 6.2 shows in more detail the collection of distillate in the flash chambers. In the last few stages, termed the "heat rejection section," cold, aerated seawater is pumped through the condensers to lower the temperature for further condensation to occur and complete the operating cycle.

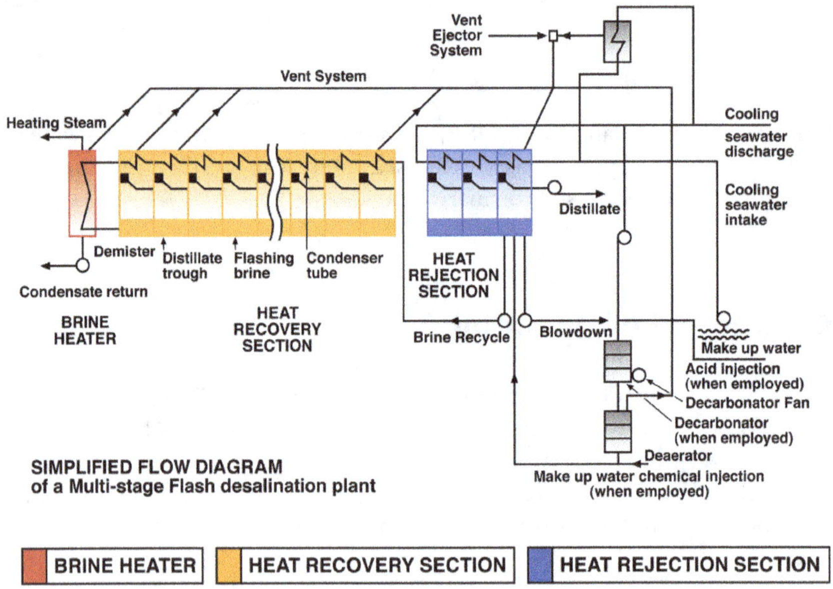

FIGURE 6.1 Schematic diagram of an MSF desalination plant.

Some of the raw seawater is discharged to waste and some is deaerated and dosed with chemicals to control scaling and foaming, before being added to the recirculating brine as feed makeup. The feed is necessary to replace evaporative losses and to maintain the recirculating brine concentration at acceptable levels. The brine is then heated in the heat recovery condensers before returning to the brine heater again. Because of the need for a supply of high temperature steam, MSF plants are usually built in conjunction with a power station.

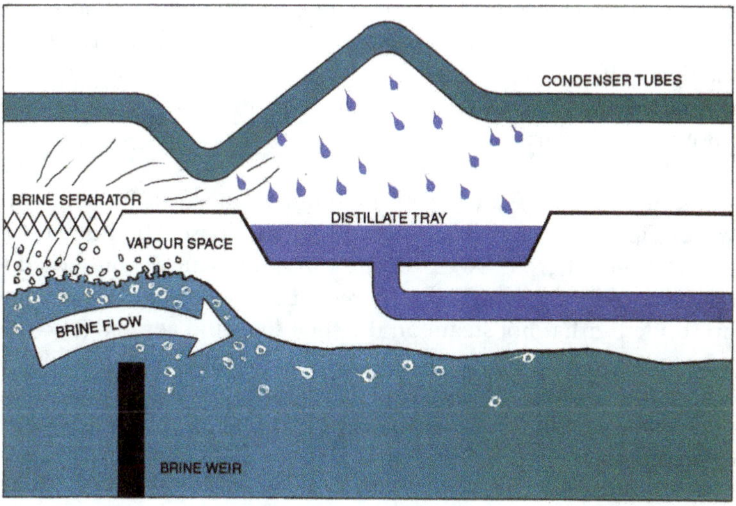

FIGURE 6.2 Detail showing collection of distillate in the flash chambers.

6.2 Conditions in MSF Plants

The conditions in the heat rejection section are typical of many seawater heat exchangers, where aerated seawater, sometimes with chlorine additions to control fouling, is on the inside of the tubes and steam is condensed at low temperatures on the shell side.

The brine fed to the brine heater and heat recovery sections is deaerated and treated with chemicals to control foaming and scaling. This is to enable carbon steel to be used for the flash chambers and to maintain heat transfer efficiency. It is usual to control dissolved oxygen to < 20 ppb, and in a well-run plant it may be < 10 ppb. From the discussion in Chapter 4, it can be seen that this keeps carbon steel corrosion rates low.

There are two principal methods of controlling scaling. One is to dose with acid to remove the bicarbonate and carbonate, which reduces the pH to around 6, so that scaling does not occur when the water is heated. The other method is to dose with proprietary antiscaling compounds. Both methods are in use, although chemical control of scaling seems to be more predominant in more modern plants.

The brine heater is also exposed to low oxygen seawater, but it also experiences the highest temperatures; a highly corrosion resistant alloy is therefore needed for the heater tubes, as described below in Section 6.3.1.

The condensate is low in chlorides and is at ambient temperature or a little higher. It is somewhat more aggressive until the pH has been finally adjusted, so 316L stainless steel or a polymer is usually used to handle it. Because the condensate is drinking water, it is essential that the polymer is approved for this purpose.

There is one further feature that is found in most MSF plants, but not shown in Figure 6.1, and this is the dump condenser. If the plant has to be shut down or there is excess steam, then the steam is condensed in the dump condenser; this is so it can be recycled to the boiler on startup. The dump condenser has hot steam on the shell side, and aerated (and possibly chlorinated) seawater on the tube side. The conditions are similar to the heat rejection section, but the steam is much hotter. The materials for the dump condenser and potential corrosion problems are discussed in the following sections.

6.3 Component Materials

The commonly used materials for many of the components in an MSF plant are discussed below. Because materials vary from plant to plant, the factors affecting the choice are discussed as well.

6.3.1 Heat Exchangers

Copper alloys are widely used for the heat exchangers in the heat recovery and brine heater sections for tubes, tube plates, and sometimes as linings for the water boxes. Aluminum brass, the copper-nickel alloys 90/10 and 70/30, and 66/30/2/2 copper-nickel-iron-manganese (Cu-Ni-Fe-Mn) alloy are all used for tubing, while both nickel aluminum bronze (NAB) and rolled naval brass are used for tube plates. Carbon steel water boxes are sometimes lined with 90/10 copper-nickel to resist corrosion

Although all four tubing alloys have been used in MSF plants, 70/30 copper-nickel has had relatively little use, while aluminum brass tends to be mostly used in the lower temperature part of the heat recovery section. The 90/10 and 66/30/2/2 alloys are currently the most widely used copper tubing alloys. The highest temperatures occur in the brine heater (~ 110 °C), and 66/30/2/2 Cu-Ni-Fe-Mn has been very successful for the tubing in this application.

The steam is hot in the dump condenser and the seawater is aerated, and the unit may be used intermittently or continuously depending on plant requirements. It is common to use 66/30/2/2 Cu-Ni-Fe-Mn for the tubes. If the unit is only used intermittently it should not be left full of stagnant seawater when not in use, because SRB could become active producing significant quantities of hydrogen sulfide (H_2S) leading to localized corrosion. The dump condenser should be drained and/or flushed with fresh water when not in use for more than 48 hours. When the dump condenser is in frequent or continuous use, then copper-nickel alloys have still been traditionally used for the tubing. However, at the time of writing, titanium tubes cost approximately the same as copper-nickel. There are rarely corrosion problems with titanium tubes, but it is important to avoid galvanic problems with copper alloys, as discussed in Chapter 4, and fatigue, as discussed below.

A survey of materials used in MSF desalination plants was carried out by Newton et al. of A.D. Little, Inc., in the 1970s.[1] More recently, in 2003, Oldfield conducted a similar survey and the results are compared in Table 6.1.[2]

TABLE 6.1 A comparison of the use of heat exchanger alloys in MSF plants from the surveys of Newton et al. and Oldfield.[1-2]

ALLOY	1972 (Ref. 1)			2003 (Ref. 2)		
	Brine Heater (%)	Heat Recovery (%)	Heat Rejection (%)	Brine Heater (%)	Heat Recovery (%)	Heat Rejection (%)
Al-Brass	17.9	68	23	0	25.7	5.6
90/10 Cu-Ni	49.0	24.5	34	27.7	52.8	18.8
70/30 Cu-Ni	1.5	1.0	1.5	7.9	11	4.2
66/30/2/2 Cu-Ni-Fe-Mn	31.3	6.0	39.5	51.8	5.2	26.5
Titanium	0.3	0.5	2.0	8.2	3.5	42.5

In Table 6.1, the materials used for heat exchanger tubes in the brine heater, heat recovery, and heat rejection sections are compared. Note that the 2003 survey divides the heat recovery section into two: the high and low temperature sections. Table 6.1 averages the data from these two subdivisions. Table 6.2 shows the difference in materials selection for these two subdivisions from the 2003 survey.[2]

The results in Table 6.1 show that the main material for the brine heater tubes has changed from 90/10 copper-nickel to 66/30/2/2 Cu-Ni-Fe-Mn. In the heat recovery section, aluminum brass has remained the most popular alloy in the cooler regions, but 90/10 copper-nickel is the predominantly used alloy in the hotter regions (Table 6.2). In the heat rejection section, there has been a big change from copper alloys to titanium. Although titanium is the most widely used single alloy, the copper alloys together still account for a majority of the tubes. One feature of interest from both surveys is the low use of 70/30 copper-nickel. The 66/30/2/2 Cu-Ni-Fe-Mn alloy appears to be regarded as a significant and useful upgrade from 90/10 copper-nickel in some parts of the plant, such as the brine heaters.

TABLE 6.2 Materials used for the heat recovery section heat exchanger tubes.[2]

ALLOY	USAGE (%)	
	Low Temperature	High Temperature
Al- Brass	46.7	4.7
90/10 Cu-Ni	47.1	58.4
70/30 Cu-Ni	0	21.0
66/30/2/2 Cu-Ni-Fe-Mn	0.4	10.0
Titanium	3.5	3.5

The Newton et al. survey also produced data on the failure rates of heat exchanger tubes; unfortunately, no such data was obtained by Oldfield. Table 6.3 shows the Newton et al. survey results.[1] These numbers need to be compared with the usage data in Table 6.1. For example, the low failure rate of 70/30 copper-nickel in the heat recovery and heat rejection sections is not surprising, given its low use. The low failure rate of 66/30/2/2 Cu-Ni-Fe-Mn is significant given the alloy's high use; 90/10 copper-nickel is also shown as a reliable alloy, as is aluminum brass in the heat recovery section.

The main reason for the increasing use of titanium in the heat rejection section is the numerous failures that have occurred with copper alloys because of sulfide polluted seawater, lodged obstructions increasing turbulence and causing erosion corrosion, and similar common causes of failure of copper alloy heat exchanger tubes in seawater.[3]

TABLE 6.3 Heat exchanger tube failure rates in MSF plants.[1]

ALLOY	FAILURE RATE (%)		
	Brine Heater	Heat Recovery	Heat Rejection
Al- Brass	15.7	1.1	6.8
90/10 Cu-Ni	2.8	0.4	2.3
70/30 Cu-Ni	15.9	0	1.6
66/30/2/2 Cu-Ni-Fe-Mn	1.5	0.02	0.05

Remember that where a heat exchanger is retubed in titanium after repeated failure of copper alloys, it will be necessary to fit additional supports for the titanium tubes on the shell side. This is because titanium does not have as high a resistance to fatigue as copper-nickel alloys, and fatigue failures have occurred where sufficient extra support was not provided.

Oldfield also reviewed the alloys used for the other parts of the heat exchangers. Table 6.4 shows the alloys used for the tube sheets in the main sections. The alloy for the dump condenser should be the same as the brine heater, because of the high steam temperatures when in use. Table 6.4 shows a high use of copper-nickel, NAB, and naval brass in the brine heater, but 90/10 copper-nickel and NAB are more common in modern plants. The tube sheet alloy usage is similar in the heat recovery section, but remember that the brine in the tubes is deaerated and less corrosive. In the heat rejection section, NAB is the most common alloy for tube sheets, with older plants using naval brass. The preference for NAB is because the seawater is aerated and very corrosive. Where the tubes are titanium, NAB is the most common tube sheet material, although accelerated corrosion of NAB coupled to titanium can occur.[4] Most heat exchangers with titanium tubes and copper alloy tube sheets have sacrificial mild steel anodes in the water boxes to protect the copper alloy. The carbon content should be $< 0.15\%$ to avoid the risk of fragmentation when the anode is corroding. Iron anodes are frequently fitted in the water boxes in many of the heat exchangers with copper alloy tubes to provide additional protection to copper alloy tube sheets. The anodes should be sized so that they only need replacement every year or two, but do not cause excessive turbulence at the tube entries.

TABLE 6.4 Materials used for the tube sheets in MSF plants.[2]

ALLOY	USAGE (%)			
	Brine Heater	Heat Recovery		Heat Rejection
		High Temp.	Low Temp.	
Naval Brass	29	30	38	33
90/10 Cu-Ni	31	28	27	6
NAB	20	22	22	50
70/30 Cu-Ni	10	11	2	5
Al-Bronze	7	6	6	6
Other	3	3	5	0

The materials used for water boxes are shown in Table 6.5, and 90/10 copper-nickel, either solid or clad onto carbon steel is the most popular in the heat recovery and brine heater stages. In the heat rejection section, rubber-lined carbon steel accounts for around a third of the capacity, with 90/10 copper-nickel being another third. Rubber-lined carbon steel works well in the heat rejection section because of the lower temperatures, but it is important that the coating is correctly applied on a properly prepared surface with appropriate curing. Water flows and temperatures in the water boxes are not high and corrosion problems in water boxes are rare, but corrosion of the carbon steel can be severe if the coating becomes damaged or detached anywhere.

The tube support plates can be made from a variety of alloys, although the mild conditions on the shell side mean that bare carbon steel is often used, as shown in Table 6.6. Alloy 316L stainless steel is also commonly used for the support plates, particularly in the heat recovery section. There is a less common use of 90/10 copper-nickel and naval brass more or less evenly across all the heat exchangers (Table 6.6). Corrosion of the support plates is unusual, but they can contribute to corrosion of the tubes in the heat recovery section under some circumstances (Section 6.6.4.2).

TABLE 6.5 Materials used for the water boxes in MSF plants.[2]

ALLOY	USAGE (%)			
	Brine Heater	Heat Recovery		Heat Rejection
		High Temp.	Low Temp.	
Rubber Coated Steel	9	0	0	31
90/10 Cu-Ni	59	84	74	34
NAB	1	0	0	17
Epoxy Coated Steel	0	0.5	10	7
70/30 Cu-Ni	16	0	0	0
Al-Bronze	5	5	5	5
Bare Carbon Steel	10	7	7	4
Other	0	0.5	1	2

TABLE 6.6 Materials used for the tube supports in MSF plants.[2]

ALLOY	USAGE (%)			
	Brine Heater	Heat Recovery		Heat Rejection
		High Temp.	Low temp.	
316L SS	35	43	43	36
90/10 Cu-Ni	13	13	13	13
Carbon Steel	45	38	38	45
Naval Brass	7	6	6	6

Oldfield summarized the data from his survey to produce a table of preferred alloys for the heat exchangers in MSF plants, as shown in Table 6.7.[9] A lesser used alternative is also shown, and plants in the survey often had mixtures of materials from either list. However, the preferred materials have shown high reliability over a period of time.

TABLE 6.7 Preferred materials for heat exchanger materials in MSF plants.[9]

SECTION	COMPONENT	ALLOY	
		Most Common	Alternative
Heat Rejection	Tubes	Titanium	66/30/2/2 Cu-Ni-Fe-Mn
	Anodes	Mild Steel	Mild Steel
	Tube Sheet	NAB	Naval Brass
	Water Boxes	90/10 Cu-Ni	Rubber lined CS
	Tube Supports	Carbon Steel	316 SS
Heat Recovery (Low Temperature)	Tubes	90/10 Cu-Ni	Aluminum Brass
	Anodes	Mild Steel	9% Ni Iron
	Tube Sheet	Naval Brass	90/10 Cu-Ni
	Water Boxes	90/10 Cu-Ni	Epoxy Coated CS
	Tube Supports	316 SS	Carbon Steel
Heat Recovery (High Temperature)	Tubes	90/10 Cu-Ni	70/30 Cu-Ni
	Anodes	Mild Steel	9% Ni Iron
	Tube Sheet	Naval Brass	90/10 Cu-Ni
	Water Boxes	90/10 Cu-Ni	90/10 Cu-Ni
	Tube Supports	316 SS	Carbon Steel
Brine Heater	Tubes	66/30/2/2 Cu-Ni-Fe-Mn	90/10 Cu-Ni
	Anodes	Mild Steel	
	Tube Sheet	90/10 Cu-Ni	Naval Brass
	Water Boxes	90/10 Cu-Ni	
	Tube Supports	Carbon Steel	Carbon Steel
	Heater Shell	Carbon Steel	

At the time of writing, the cost of titanium grade 2 heat exchanger tubing was only a little more expensive than aluminum brass, and of a similar price to 90/10 copper-nickel. Hence, there is no significant cost penalty in using titanium in the heat exchangers with the most aggressive conditions, i.e., the heat reject section and the dump condenser. Titanium can also be considered for the brine heater. This is not a problem in new heat exchangers, which can be designed with extra support plates to prevent fatigue of the titanium tubes. In re-tube situations, it is necessary to provide intertube supports to reduce vibration.

The cost of duplex stainless steels has also decreased considerably in recent years compared with austenitic alloys. This means that 2304 lean duplex can be used as a lower cost alternative to 316L stainless steel for tube sheets, baffle plates, and tube support plates in the heat recovery section and for baffles and tube support plates in the heat rejection section. Duplex 2205 can be used cost effectively for tube sheets and water boxes with titanium tubes, if mild steel sacrificial anodes are fitted in the water boxes. This is already being done very successfully in multiple-effect distillation (MED) desalination plants as described in Chapter 7.

6.3.2 Pumps

The positions of the main pumps in an MSF plant are shown in Figure 6.3, along with the pumped fluid, temperature, and degree of aeration. Some, but not all of these pumps, were covered in the survey of materials' use in MSF plants by Oldfield.[2]

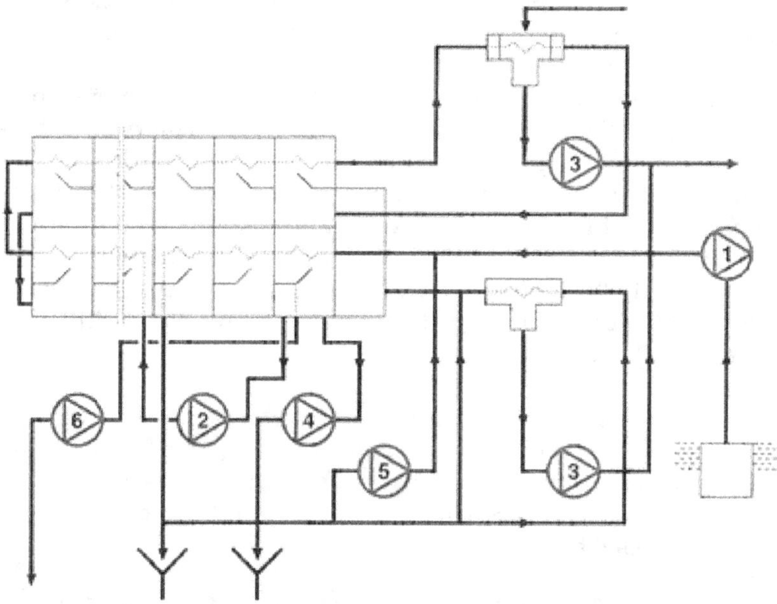

FIGURE 6.3 The main pumps in an MSF desalination plant.

1. Seawater supply, 39 °C; aerated
2. Brine recirculation, 60 °C; deaerated
3. Brine Heater, 110 °C; deaerated
4. Brine blowdown, 40 °C; deaerated
5. Seawater recirculation, 33 °C; aerated
6. Produced water, 31 °C; deaerated

The seawater supply pumps and seawater recirculation pumps both handle warm, aerated seawater and must be in corrosion resistant alloys. Historically, pumps for both applications had austenitic cast iron (ACI) cases (and bell mouth), with stainless steel impellers and shafts. The stainless steel was mostly either 316 or 2205 duplex. Neither of these grades is resistant to crevice corrosion in seawater, but they have been satisfactory in service because of the galvanic protection provided by the case. However, as explained in Chapter 4, there have been numerous failures of austenitic cast iron due to stress corrosion cracking (SCC), particularly in warmer waters where MSF plants are often situated. The trend has been to replace these pumps with ones in superduplex stainless steel, which has a history of successful use in warm seawater. Although cast 2507 and cast Z100 have both been used, cast Z100 is preferred for larger pumps because it is easier to cast in thick sections, with less risk of cracking.[5]

The column pipes have mostly been in ACI when the case is ACI, although 316L and NAB have also been used. When the pump is superduplex stainless steel, the column pipes should also be superduplex stainless steel for galvanic compatibility.

The brine recirculation pump operates at higher temperatures, but the water is less corrosive because of the low oxygen content (< 20 ppb). Austenitic cast iron is commonly used for the pump case, although cast iron has been used in some older plants. ACI generally works well in this application because of the low oxygen content of the brine. The impellers are mostly (63%) 316 stainless steel, although 2205 duplex stainless steel has also been used (23%). Vertical pumps have often used either an fiberglass reinforced plastic (FRP) or coated carbon steel barrel (60%), with 316 stainless steel, NAB, and ACI being used less.[2]

The brine blowdown pumps are similar to the brine recirculation pumps, but operate at a somewhat lower temperature (Figure 6.3). The materials of construction are therefore the same, with similar usage rates for the different components, although 2205 duplex stainless steel is more widely used in horizontal blowdown pumps.[2]

The conditions in the produced water pumps are not particularly aggressive, because of the low chloride and oxygen concentrations. The stainless steel 316 is the most common alloy for both impellers and shafts (97%) in vertical pumps, although NAB and 2205 have also been used in horizontal pumps. The shafts are usually 316 in horizontal pumps. The barrel in vertical pumps is often FRP (38%), but 316 stainless steel (21%), aluminum bronze (14%), and coated carbon steel (18%) have all been used. In horizontal pumps, the case may be in a variety of materials including NAB (32%), 316 stainless steel (32%), and ACI (11%).[2]

The brine heater pump operates in conditions similar to the brine recirculation pumps, but at significantly higher temperatures. Although notionally deaerated, there will be some oxygen present and at these temperatures, there is a risk of SCC of 316 stainless steel, although it can work if dissolved oxygen control is good. However, cast duplex stainless steel has much better resistance to SCC and cast 2205 is not much more expensive, because of its much lower nickel content than the 316 grade. Hence, 2205 would be preferred for the impellers and cases. Wrought 2205 can be used for the shaft.

6.3.3 Piping

Piping is used for a wide range of purposes in an MSF plant; the seawater intake is discussed in the next section, while this section deals with the rest of the system. The results of Oldfield's survey are summarized in Table 6.8.[2]

TABLE 6.8 Materials used for piping from Ref. 2.

ALLOY	USAGE (%)		
	Seawater Piping	Brine Piping	Produced Water
FRP	46	0	0
90/10 Cu-Ni	28	73	0
Epoxy Coated Carbon Steel	20	7	0
Rubber Lined Carbon Steel	5	0	0
Flake Glass Lined Steel	0	9	0
Carbon Steel	1	4	5
316L	0	7	85
304L	0	0	10

It can be seen that FRP is the material most used for seawater piping, but with significant use of 90/10 copper-nickel and epoxy coated carbon steel. FRP is most cost effective in larger diameters, while 90/10 copper-nickel is most cost effective in smaller sizes. The whole issue of cost effective selection of piping is complex and project specific. The relative costs of different piping materials have been investigated for a number of projects, e.g., Refs. 7 and 8. It is important to consider not only the cost of the pipes, fittings, and flanges, but also supports, fabrication, etc. One thing that both studies showed is that different materials become lowest cost over different size ranges. Corrosion resistant alloys are usually cheapest in smaller sizes, rubber-lined steel in mid-range sizes, and FRP in large sizes. The crossover point is project specific and depends on the quantities of specific items, such as valves, as well as any local requirements. The requirements for an MSF desalination plant, for example, are very different to the oil and gas applications discussed in Refs. 7 and 8. In order to get an idea of which are the most cost effective materials in specific sizes, it is necessary to do a proper evaluation for the specific project.

It can be seen in Table 6.8 that 90/10 copper-nickel is by far the most widely used alloy for brine piping and it has an excellent track record in service. There are no known specific problems with the other materials in Table 6.7, although 316L could suffer SCC if used above 60 °C. This is unlikely inside the pipe if the oxygen content is low, but may occur on the outside if condensation or seawater splashing occurred. A lower cost alternative would be 2304 duplex, which also has much better resistance to chloride SCC.

The product water piping was entirely in stainless steel in Oldfield's survey, with 316L being the most common alloy.[2] Alloy 304L stainless steel is acceptable where temperatures and chloride contents are lower. A lower cost alternative to 316L could be 2304 duplex if the fittings and flanges are readily available.

There are some other applications for piping in an MSF plant. Where acid cleaning is carried out, the piping is almost always 316L. Where acid dosing is carried out to prevent scaling, the piping is usually rubber lined or polypropylene-lined carbon steel, although other alloys such as 316L could be considered, depending on the acid and its concentration. Chemical antiscalants are usually handled in 316L stainless steel piping as are antifoam additions.

6.3.4 Seawater Intake

The seawater intake is often a long, concrete-lined channel or concrete piping that may be a kilometer or more long, because it takes seawater from a reasonable depth into a basin where the seawater circulation pumps are installed. Where seawater is pumped into the basin, the piping is usually FRP. The seawater must be filtered to remove debris and this can take the form of static screens or rotating band screens, depending on the location and size of the plant.

Traditionally, such screens were often in 316L stainless steel with steel or cast iron frames, and the 316L worked because it was galvanically protected by the carbon steel/cast iron. In modern plants, the steel would usually be coated, or the frame or body may be a polymer and the 316L will then be susceptible to crevice corrosion. This can be avoided by the use of superduplex stainless steel, which is strong as well as corrosion resistant, and it is readily available as wire, mesh, or other thin sections required for screens and filters.

Where chlorine or hypochlorite is added to control fouling, this may be at the intake entrance, especially if it is long, and/or into the seawater basin. It is important that the concentrated chlorine or hypochlorite is injected in such a way that it dilutes quickly in the seawater before it contacts any metal components that might otherwise suffer accelerated corrosion.

6.3.5 Valves

The materials for valves will be discussed according to the fluid inside, i.e., aerated seawater, deaerated hot brine, or produced water.

Almost all valves in the aerated seawater system are butterfly valves, with the most common materials now being a rubber-coated steel body with an NAB disc. A number of other materials have been used for the shaft in the past, such as alloy 400 or K-500, but these days it tends to be stainless steel, either 316L or 2205 duplex. If it is 316L or 2205 duplex, their success depends on galvanic protection from the NAB. The use of a superduplex stainless steel shaft will not cause galvanic corrosion of the NAB because the area of the disc is much greater than the wetted area of the shaft.

If there are sulfides in the seawater, then severe corrosion of the NAB disc can occur and it will not seal. In such cases, the best choice is still a rubber-coated steel body, but with a superduplex stainless steel disc and shaft.

In the deaerated brine, the oxygen content is low, but the temperatures are high and the materials selection changes. The majority of the valves are still butterfly, but the body is bronze (NAB in modern plants), although uncoated carbon steel and ductile cast iron have also been used to a lesser extent.[2] The disc is also bronze (NAB in modern plants), while the shaft can be a variety of alloys. In older plants, the shaft was often alloy 400, but this is expensive and nowadays stainless steel is more common. Duplex 2205 is preferred, but 316L has also been used in some plants.

Produced water valves are also mainly of the butterfly type, with predominantly 316L discs and shafts. The body is also mainly 316L, although older plants have used cast iron bodies.

In the vapor lines, butterfly valves again predominate with 316 discs and a stainless steel shaft. Although 316L has been used for the shaft, so have 13% Cr martensitic alloys, because the vapor is not very corrosive and this alloy has generally proved to be satisfactory. The body is generally in 316 stainless steel, although cast iron bodies have been used in a few plants.[2]

6.3.6 Flash Chambers

The flash chambers consist not only of the shell, but also the ancillary equipment, which includes the weirs, or nozzles, the demister pads, and the demister support frames. In Oldfield's survey of MSF plant materials, those used in the high- and low temperature heat recovery sections were often different.[2] However, the definition of high and low temperature varied, with some responses just calling the first two stages "high temperature," while in other responses, the high temperature section was called "half the heat recovery" stages. No distinction is made between these two in this discussion, which just refers to the high and low temperature stages.

The materials selection can be divided into the brine side and the vapor side. Oldfield condensed his survey data into a table showing the preferred and alternative material selections for the different components (Table 6.9).[9]

It can be seen that carbon steel is widely used apart from the shell in the low temperature section, but that 316L is the most widely used alloy in the high temperature section. Although not shown in Table 6.9, as an alternative to 316L for the demister pads and support frames on the brine side, some plants have used the titanium stabilized version of the alloy (UNS S31635). This has no significance from a corrosion point of view, and both versions are fully weldable.

With the reduced cost of duplex stainless steels compared with austenitic alloys as discussed in earlier sections of this chapter, it can be cost effective to use lean duplex instead of 316L for flash chamber shells and other components listed in Table 6.9 (see Section 6.5). Lean duplex stainless steels are already widely used for evaporator shells in modern MED plants (see Chapter 7).

TABLE 6.9 Standard material selection for the flash chambers and ancillary equipment in MSF plants.[9]

SECTION	COMPONENT	ALLOY	
		Most Common	Alternative
Brine Side: High Temperature	Shell	Carbon Steel	Carbon Steel
	Lining/ Coating	316L SS	90/10 Cu-Ni
	Floor	316L SS	90/10 Cu-Ni
	Side Walls	316L SS	90/10 Cu-Ni
	Interstage Walls	316L SS	90/10 Cu-Ni
	Roof	316L SS	90/10 Cu-Ni
	Weirs or Nozzles	316L SS	66/30/2/2 Cu-Ni-Fe-Mn
	Demister Pads	316L SS	Alloy 400
	Support Frames	316L SS	90/10 Cu-Ni
Brine Side: Low Temperature	Shell	Carbon Steel	Carbon Steel
	Lining/ Coating	Epoxy Coated CS	None
	Floor	Epoxy Coated CS	Carbon Steel
	Side Walls	Carbon Steel	Epoxy Coated CS
	Interstage Walls	Carbon Steel	Epoxy Coated CS
	Roof	Carbon Steel	Epoxy Coated CS
	Weirs or Nozzles	Carbon Steel	Epoxy Coated CS
	Demister Pads	316L SS	Alloy 400
	Support Frames	316L SS	90/10 Cu-Ni
Vapor Side: High Temperature	Shell	Carbon Steel	Carbon Steel
	Lining/ Coating	316L SS	None
	Floor	316L SS	Carbon Steel
	Side Walls	316L SS	Carbon Steel
	Interstage Walls	316L SS	90/10 Cu-Ni
	Roof	316L SS	Carbon Steel
	Distillate Trays	316L SS	90/10 Cu-Ni
	Dist. Transfer Trough	316L SS	90/10 Cu-Ni
Vapor Side: Low Temperature	Shell	Carbon Steel	Carbon Steel
	Lining/ Coating	316L SS	None
	Floor	Carbon Steel	316L SS
	Side Walls	Carbon Steel	316L SS
	Interstage Walls	316L SS	Epoxy Coated CS
	Roof	Carbon steel	316L SS
	Distillate Trays	316L SS	90/10 Cu-Ni
	Dist. Transfer Trough	316L SS	90/10 Cu-Ni

6.3.7 Vent System

The vent system handles the incondensable gases from the flash chambers and elsewhere in the plant. The most corrosive of the incondensable gases is carbon dioxide, which lowers the pH and increases the risk of corrosion. It consists of various piping lines from the brine heater and flash chambers, either venting directly or cascading down chambers prior to venting. The vented gases then pass through condensers and finally to an ejector. There are often three condensers termed the "after condenser" (hottest, ~ 170 °C), the "intercondenser," and the "vent condenser" (coolest). The condensers may be shell and tube type or direct contact.

The most common venting system material for the vent piping and ejectors is 316L stainless steel.[2] The after condenser has the most aggressive conditions and if it is of the shell and tube type, then the tubes are usually titanium with 70/30 copper-nickel tube sheets, and 316L stainless steel or 90/10 copper-nickel–clad steel water boxes. Where the condensers are the direct contact type, the shell may be 316L stainless steel, but the aggressive conditions mean that titanium and alloy 825 (UNS N08825) are also widely used, with 316L or alloy 825 for the intervessel piping.[2,9]

Where 316L has been used, and the conditions are particularly aggressive so that excessive corrosion or SCC has occurred, then replacement with 2205 duplex stainless steel is a lower cost alternative to alloy 825. Alloy 2205 has better resistance to localized attack by acidic chlorides than 316L and it also has excellent resistance to chloride SCC (see Chapter 4).

The ejector condensers handle wet incondensable gases at temperatures up to 170 °C, where carbon dioxide is the main incondensable gas. Oldfield and Todd described how alloy 316L stainless steel can handle wet CO_2, but also how problems can arise in some plants.[10] They also described what happens when chlorine or hypochlorite is added to control fouling in seawater. The chlorine reacts with the bromide in seawater to form bromine and hypobromous acid. In plants where acid is dosed to remove carbonates and bicarbonates, the principal product at low pH (< 5) is bromine. Even at pH from 5 to 6 there is still a significant bromine concentration. This is then stripped out in the venting system where it concentrates by 1,000 to 10,000 times.[10] The bromine gas is very corrosive both in the liquid and particularly in the vapor phases.[11] Corrosion of 316L will occur as pitting in numerous places, which quickly perforate. This can occur at concentrations of bromine in the vapor exceeding 10 mg/L.[10] Oldfield and Todd showed the minimum pH to avoid corrosion of 316L at different chlorine concentrations at the entry to the deaerator.[10]

This problem can also occur in both acid-dosed and scaling-inhibitor dosed plants if there are small amounts of ammonia present as a pollutant in the seawater.[10] As little as 0.3–0.6 mg/L ammonia is sufficient to form significant quantities of bromamine where chlorine or hypochlorite is added to the seawater. In the vent system, the bromamine is stripped from the seawater where it decomposes back to ammonia and bromine, creating a very corrosive environment. This has led to failures of 316L venting systems.[10]

Hodgkiess et al. looked at the corrosion of materials in wet bromine vapor and found that only titanium (grade 2) and alloy 28 (UNS N08028) had significant resistance to bromine.[11] To avoid the use of high cost materials in the ejector condenser, one solution is to inject water to lower the temperature from 170 °C to ~ 130 °C. At this temperature, it is possible to use 904L, 6% Mo austenitic, or superduplex stainless steels. Alloy 904L is the lowest alloyed of these materials, but it is not produced in the same quantities as the other alloys, and it contains 25% nickel, so the alternatives can be cheaper and more readily available.

6.3.8 Deaerators and Decarbonators

The deaerator is mostly carbon steel lined with 316L, with 316L stainless steel nozzles and piping.[9] The decarbonator (when fitted) is usually made from rubber-lined carbon steel, also with 316L stainless steel nozzles and piping.[9]

6.3.9 Instrumentation Tubing

Instrumentation lines are usually small in diameter, typically 10 or 12 mm (¾ or ½ in.), while the main seawater/brine piping can be nominal pipe size (NPS) 6 or larger. The selection of materials for instrumentation tubing from piping carrying aerated seawater was discussed in Chapter 5 for seawater reverse osmosis (SWRO) plants, and the conclusions apply equally to instrumentation tubing for aerated seawater lines in MSF plants.

Figure 6.4 is a schematic diagram showing the arrangement for an instrument tubing takeoff, where the water in the tubing is often static, such as to pressure gauges. Although a wide range of alloys is available for tubing, in *deaerated* brine lines it is satisfactory to use 316L stainless steel or 90/10 copper-nickel, and other, more expensive alloys are not required.

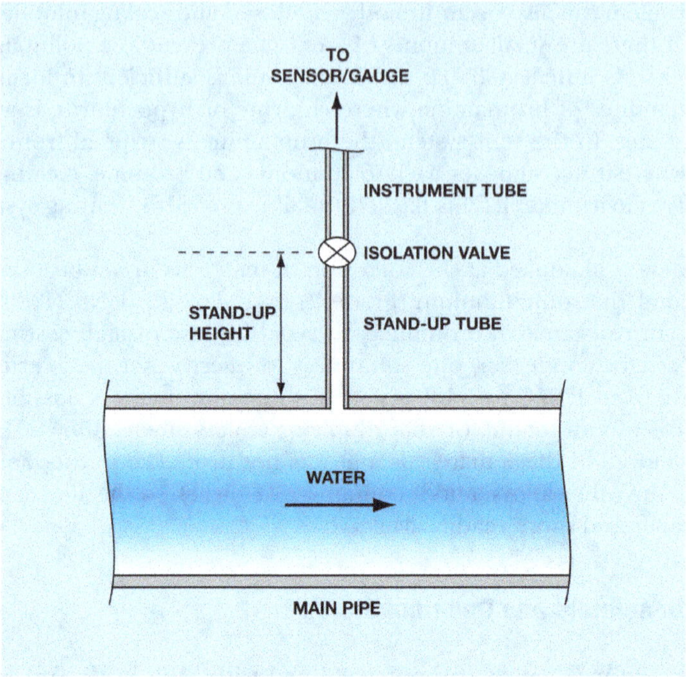

FIGURE 6.4 Diagram showing arrangement of takeoff for instrument tubing.

With deaerated seawater, where the dissolved oxygen is ≤ 20 ppb, there are no problems with corrosion of 316L or 90/10 copper-nickel internally up to at least 110 °C. However, where the 316L tubing has an external temperature ≥ 60 °C, there is a risk of chloride SCC if water is splashed outside of the tubing. Water with chlorides as low as 10 mg/L may cause SCC of 316L, because the chlorides concentrate by evaporation. Hence, 90/10 copper-nickel is preferred for all applications where the instrumentation tubing temperature will exceed 60 °C.

Oxygen excursions up to 50 ppb or more in the main line should have no significant effect on corrosion of 90/10 copper-nickel tubing. There is no data on the corrosion of 90/10 copper-nickel at temperatures exceeding 110 °C with low oxygen contents.

Whichever alloy is chosen, there is no need for a minimum stand-up height. The choice of isolation valve material is governed by temperature. Both 316L and NAB are satisfactory from an internal corrosion perspective, but there is a risk of external SCC of 316L at temperatures above 60 °C. At lower temperatures there is no risk of SCC of 316L and so there is no need for external coating. At higher temperatures a coating is desirable. However, it is difficult to find coatings that are reliable at temperatures above ~ 70 °C. At temperatures of 70 °C and above, an NAB valve and 90/10 copper-nickel tubing would be better, because they are not susceptible to chloride SCC. External crevice corrosion, at clamps for example, can occur when 316L tubing is splashed with seawater. A coating will reduce the risk of external corrosion or 90/10 copper-nickel can be used instead.

All the comments in this section also apply to tubing carrying flowing, deaerated water, to analyzers, etc., that is, the low dissolved oxygen content means that either 316L or 90/10 copper-nickel will be satisfactory and higher alloy tubing is not necessary.

6.3.10 Produced Water Treatment

The produced water is very pure with low chlorides and no significant quantities of other ions. It will not have a stable pH so it is common to take the CO_2 from the vent system and add it to the produced water, or some of the produced water, to lower its pH. The water is then passed through a column of limestone where the water and CO_2 react with the limestone to produce calcium bicarbonate:

$$CO_2 + H_2O + CaCO_3 \rightarrow Ca(HCO_3)_2.$$

The calcium bicarbonate will buffer the pH, and as little as 50 mg/L bicarbonate can have a significant buffering action. The piping is usually 316L stainless steel and the reaction columns are either carbon steel lined with rubber or 316L stainless steel.

As discussed above, 2304 duplex is lower cost than 316L and can be considered for the piping and vessels.

The water is then adjusted to a pH in the range 7 to 8, usually by additions of lime, and some seawater is often added to provide some salts in the water. Finally, the water is chlorinated to disinfect it so that it can go into the potable water distribution system.

6.3.11 Ancillary Equipment

Materials for chemical storage tanks and chemical injection pipework are discussed in Chapter 5, Section 5.3.4 for SWRO plants, and the same material selections are recommended for such tanks, and are applicable to such tanks and pipework in an MSF plant.

There are many items of equipment that are not exposed to aggressive environments internally, but are exposed to the atmosphere. Such equipment includes electrical control and junction boxes, storage tanks, etc. These may be made of carbon steel, low-alloy stainless steel (304L or 316L), or FRP. The atmosphere at many MSF plants is both hot and laden with salt from proximity to the sea. This means it will be classed as C5 or CX, the most aggressive atmospheres, by ISO 9223.[12] Hence, carbon steel will require a high quality paint coating and low-alloy stainless steel will require regular cleaning, and possibly painting. This is discussed in detail in Chapter 4, Sections 4.1 and 4.4 for carbon steel and stainless steel, respectively. The protection of FRP enclosures from UV is discussed in Chapter 2, Section 2.5.3.

6.4 Service Experience

There have been numerous corrosion problems related to MSF plants published over the last 50 years. In this section, some corrosion failures observed by the author in the last few years are reported.

6.4.1 Heat Rejection

Chlorinated, aerated seawater is aggressive and if sulfides are also present, the mixture can be very corrosive to copper alloys.[3] An MSF plant in the Middle East had tubes of 66/30/2/2 Cu-Ni-Fe-Mn in the heat rejection section, which corroded very rapidly during the first few years of operation. Some of the heat exchangers were re-tubed because of the severity of the damage, as shown in Figure 6.5. There were lots of pits filled with a yellow-green product, particularly in the center of the tubes. The corrosion products gave a strong reaction for sulfide in the sodium azide test,[13] and research has shown that the 66/30/2/2 Cu-Ni-Fe-Mn alloy is the most susceptible of the copper alloys used for heat exchanger tubes to sulfide attack in the presence of chlorine (see Chapter 4).[3]

(A)

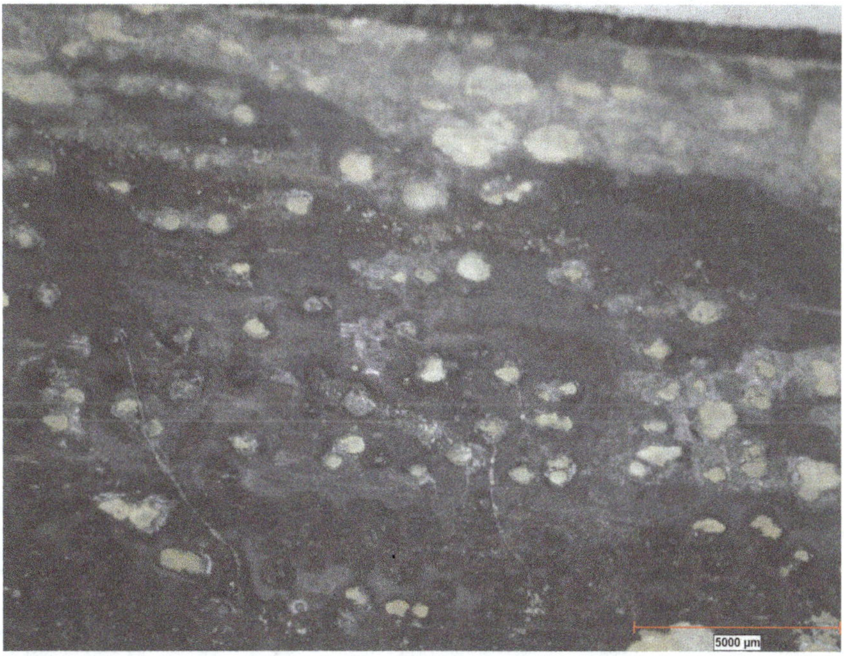

(B)

FIGURE 6.5 Appearance of 66/30/2/2 Cu-Ni-Fe-Mn tubes after attack by sulfides: A) general appearance, and B) detail of pitting.

The seawater feed pipes to the heat rejection sections were large-diameter FRP with a total length of several hundred meters. There had been numerous extended shutdowns for maintenance, and the feed pipes had been left full of stagnant water. During this time, the aerobic bacteria would consume the oxygen and the SRB could then become active. After a shutdown of 15 days, 15 mg/L sulfide was measured in the seawater feed piping. As little as 0.01 mg/L sulfide can cause accelerated attack of 66/30/2/2 Cu-Ni-Fe-Mn.[3] It was suggested that during shutdowns exceeding 48 hours, the vent pipe near the condenser inlet should be directed to waste so that a slow continuous flow of seawater was maintained through the feed pipe. The alternative to this was a complete re-tube of the heat rejection section in titanium, which would require additional changes to the heat exchanger to maintain satisfactory operation.

In the same plant, there were some butterfly valves with rubber-lined steel bodies and NAB discs, which had suffered severe corrosion, as shown in Figure 6.6. NAB is also very susceptible to attack by sulfides (see Chapter 4) and a change to superduplex stainless steel discs was suggested.

FIGURE 6.6 Severe corrosion at edge of NAB butterfly valve disc due to sulfides.

The author has also seen erosion corrosion failures of 90/10 copper-nickel tubes due to lodged obstructions, when the seawater filters have been damaged.

Some plants have suffered pitting failure of 90/10 copper-nickel tubes when there is ammonia pollution in the incoming seawater from 0.3–10 mg/L. In one case, this was prevented by dosing with ferrous sulfate solution for

1 hour per day for 60 days, until protective films had formed. Dosing was discontinued after 60 days, when stable protective films had formed, and no pitting occurred subsequently.

A second plant with regular ammonia concentrations in the cooling water was tubed with aluminum brass, because an earlier plant had suffered severe corrosion of its 90/10 copper-nickel tubes. Research had shown aluminum brass to be more resistant to ammonia pitting.[3] Ferrous sulfate dosing was carried out every day with the aluminum brass tubes because not only did it help to resist ammonia attack, it also improved resistance to erosion corrosion.[3]

6.4.2 Heat Recovery

Oldfield and Todd reviewed the causes of corrosion on the vapor side in the heat recovery section.[13] The corrosion is principally caused by oxygen or carbon dioxide. Oldfield and Todd demonstrated that air leaking past both gaskets and sight glasses in the heat recovery section could be sufficient to cause corrosion of carbon steel. They found corrosion was most severe on the interstage walls and was often worst above the distillate transfer trough. Corrosion rates were up to 1 mm/y. Corrosion resistant alloys in the vapor space were unaffected. Oldfield and Todd also found that corrosion due to oxygen ingress was often worst in the middle stages of the heat recovery section.[13]

Carbon dioxide comes from the bicarbonate ion that is always present in seawater, and this will pass into the deaerated brine in plants using additives to control scaling. Most of the carbon dioxide is released in the high temperature stages. However, it is common practice to cascade incondensable gases from stage to stage and just vent them off in a few stages. If the venting is only from the mid and lower stages, the carbon dioxide concentration in the hotter sections can be significant. The effect of carbon dioxide is to lower the pH below the stability range of cuprous oxide, so that copper alloys cannot form a protective film ($pH < 6$). This low pH can also cause increased corrosion of carbon steel in the flash chambers.[14] The answer is to use more vents and less cascading, to prevent the buildup of CO_2 concentrations.

The author examined some severe external corrosion of 90/10 copper-nickel tubes from the high temperature heat recovery section of an MSF plant. The corrosion took the form of erosion corrosion, which was particularly severe at the tube support plates (Figure 6.7). It was assumed that the CO_2 concentration was very high so a protective film could not form on the tubes. The fact that the vapor was in vigorous motion around the tubes would cause the erosion corrosion, when it would not occur at near neutral pH. The tube support plates were 316L stainless steel, and the exacerbated corrosion at the plate/tube junction could be due to acceleration of the vapor through the plate/tube gap, or the corrosion may have been accelerated by galvanic action, with the 316L as the cathode, or a combination of both mechanisms.

FIGURE 6.7 Severe external corrosion of 90/10 copper-nickel at tube support plate in heat recovery section.

An inspection of the 316L vent system showed that it was badly corroded and cracked, and was not working correctly. This explained the reason for the CO_2 buildup. It was suggested that the vent system be repaired and 2205 duplex stainless steel be considered for sections suffering from SCC.

6.4.3 Dump Condenser

Dump condensers are often high up so that they can use the siphon effect to increase water flows and reduce pumping demand. However, this can give rise to corrosion under some circumstances.

The author examined some 66/30/2/2 Cu-Ni-Fe-Mn tubes from the dump condenser in an MSF plant and observed severe erosion corrosion in some of the tubes, as shown in Figure 6.8. All the attack was near the outlet and towards the bottom of the condenser, which is typical of syphonic air release, as shown schematically in Figure 6.9. Gas is released under syphonic conditions, and the turbulence caused by this can result in severe erosion corrosion, even for an alloy like 66/30/2/2 Cu-Ni-Fe-Mn, which is very resistant to erosion corrosion.

FIGURE 6.8 Severe erosion corrosion of 66/30/2/2 Cu-Ni-Fe-Mn tube from a dump condenser.

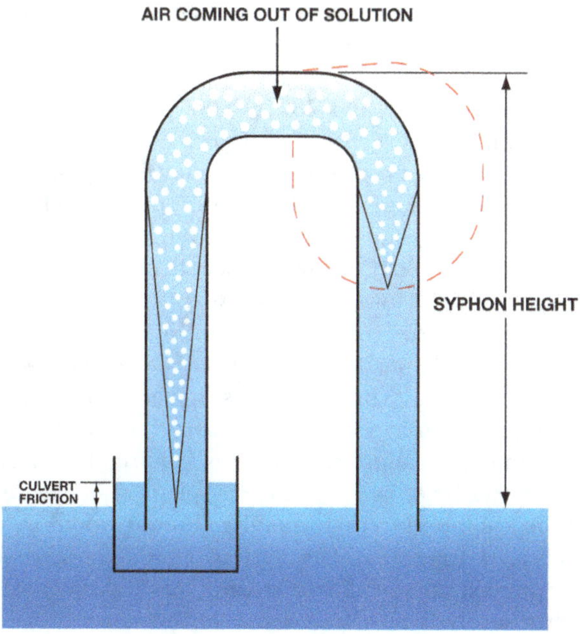

FIGURE 6.9 Schematic diagram showing release of air under syphonic conditions.

An analysis of real condensers shows that the inlet and outlet sections of the second pass are most susceptible to attack.[15] Syphonic air release does not occur if the pressure in the tubing is maintained above 0.6 bar.[15]

The solution was to adjust the valves in the seawater system to maintain an adequate pressure in the condenser tubes. An eddy current survey showed just a few tubes near the bottom were severely attacked and these were plugged. Since then there have been no further problems.

6.5 Future Trends

There are a number of technical advances that could affect materials in MSF plants and the corrosion performance.

In MSF plants, the present maximum operating temperature (~ 110 °C) is limited by the scaling tendency at higher temperatures, even with scale control chemical additions. One method being investigated to overcome this limit is the use of nanofiltration. This removes all particles down to 0.001 μm and effectively removes the ions that lead to scaling. If this can be achieved reliably and at a reasonable cost, then higher brine temperatures are possible.

With brine temperatures of 160–180 °C MSF plants would be more efficient. The question then is what materials to use for the tubes, tube plates, and water boxes? The most commonly used alloy for the brine heater tubes is 66/30/2/2 Cu-Ni-Fe-Mn followed by 90/10 copper-nickel. Unfortunately, there is no data to show what the corrosion rates of these alloys might be at temperatures of 160–180 °C, or what effect dissolved oxygen might have in the 10–50 ppb range. However, in the years immediately after the Second World War, many power stations in the United Kingdom used 70/30 copper-nickel tubes in the steam heaters. These performed well at temperatures up to about 190 °C. When stations switched from continuous to intermittent operation, exfoliation corrosion sometimes became a problem.[16] More research and application experience is required to determine whether copper alloys will be satisfactory in MSF plant at higher temperatures. If copper alloys are found to be unsuitable, then it would be necessary to consider alternative materials. The obvious candidate alloys are 22% Cr duplex, superduplex, and 6% Mo austenitic stainless steels and titanium. Unfortunately, there was no data on the performance of these materials under the proposed operating conditions at the time of writing.

Current practice in MSF plants is to use carbon steel flash chambers lined with 316L in the regions most affected by corrosion. In addition, the external surfaces require regular painting. In 1992, Olsson and Groth examined the economics of solid stainless steel evaporators compared with carbon steel.[17] The duplex stainless steels are the most attractive because their high strength gives the biggest reduction in wall thickness compared to carbon steel. External stiffeners are required to utilize the maximum wall thickness reduction, but with widely spaced, profiled stiffeners the cost will be low.[17] Their analysis showed that over the full 20 to 25-year life of a plant, savings of ~ 20% were possible by using solid 2304 or 22% Cr duplex stainless steel instead of 316L clad carbon steel, which required regular external painting. The cost analysis from 1992 is shown in Table 6.10, and the lifecycle cost benefit is probably better using current prices.

TABLE 6.10 Cost calculated in 1992 in U.S. dollars (USD), for the fabrication of a simplified flash chamber shell in carbon steel and stainless steel.[17]

	ALLOY			
	C-Steel	**2304**	**2205**	**316**
Material				
Weight (kg)	26,830	14,054	13,141	19,469
USD/kg	0.5	2.80	3.25	2.65
Cost (USD)	13,415	39,351	42,708	51,590
Fabrication	48,000	38,700	38,700	38,700
Lining				
Area (m^2)	117			
USD/m^2	200			
Cost (USD)	23,400			
Painting				
Area internal (m^2)	117			
USD/m^2	45			
Area external (m^2)	234			
USD/m^2	25			
Painting Cost (USD)	11,115			
Total Cost (USD)	95,930	78,051	81,408	90,290

More recently, Olsson et al. described the use of 22% Cr duplex stainless steel flash chambers at the Wafa MSF plant in Libya.[18] Although the plant's designers did not take full advantage of all the savings possible, i.e., by not using the lowest cost alloy (2304) for the flash chambers, the operators of the plant still expect to save on operating costs over the life of the plant, compared with if 316L clad carbon steel had been used throughout.

One possibility that has been suggested is a once-through plant using fully aerated seawater. While this offers savings in equipment, such as deaerators and brine recirculation pumps, it requires high-alloy materials for all the heat exchangers and flash chambers. A single small plant has been built using this concept at Sirte in Libya, using 6% Mo austenitic stainless steel for the most part. However, for a large plant, it has been argued that the increased cost of the heat exchangers and flash chambers, more than offsets the savings at current prices, compared to a conventional MSF plant.[19]

Finally, the discussion above has shown that 316L is commonly used in MSF plants for a wide range of applications. However, experience has shown that in some instances the conditions are aggressive enough to cause corrosion and/or SCC. A cost-effective alternative would be lean duplex for increased SCC resistance, and 2205 duplex where extra corrosion resistance is required as well. Even where SCC is not a problem, 2304 is a lower cost alternative to 316L, where it is available in the appropriate product forms.

REFERENCES

1. Newton, E.H., J.D. Barckett, and J.M. Ketteringham, "Survey of Materials in Large Desalting Plants Around the World," through A.D. Little, Inc. (Washington, DC: U.S. Office of Saline Water, 1972).
2. Oldfield, J.W., "Novel Material Selection to Improve Corrosion Resistance," Report to the Middle East Desalination Research Centre, Oman, Project 97-AS-001 (Hastings, United Kingdom: Edgerton Publishing Services, 2003).
3. Francis, R., *The Corrosion of Copper and Its Alloys: A Practical Guide for Engineers* (Houston, TX: NACE International, 2010).
4. Francis, R., *Galvanic Corrosion: A Practical Guide for Engineers*, 2nd ed. (Houston, TX: NACE International, 2017).
5. Francis, R. and S. Hebdon, "The Selection of Stainless Steels for Seawater Pumps," CORROSION 2015, paper no. 5446 (Houston, TX: NACE International, 2015).
6. Francis, R., *The Selection of Materials for Seawater Cooling Systems: A Practical Guide for Engineers* (Houston, TX: NACE International, 2006).

7. Grimm, G.C., and A. Bliall, "The Use of GRP Piping in the Oil and Gas Ballast Water System of the Draugen Gravity-Based Structure," Conference on Polymers in the Marine Environment (London, United Kingdom: Institute of Marine Engineering, 1991).
8. Onsum, A.S., "*Kosthadseffective Materialvalg*" (Cost Effective Material Selection) Conference, "Materialteknologi i oljeindustrrein" (Materials Technology in the Oil Industry) (Oslo, Norway: Norwegian Society of Engineers and Technologists, 2002).
9. Oldfield, J.W., "Survey of Material Usage in MSF Plants over the Past 25 Years," IDA Conference (Topsfield, MA: International Desalination Association, 2003).
10. Oldfield, J.W., and B. Todd, "Corrosion of Stainless Steels Caused by Bromine Emission from Chlorinated Seawater," Marine Corrosion of Stainless Steels: Chlorination and Microbial Effects, European Federation of Corrosion Publication No. 10 (London, United Kingdom: Institute of Materials, 1993): p. 92.
11. Hodgkiess, T., N.K. Ng, and P. Argyropoulos, *The Corrosion Behaviour of a Number of Materials Exposed to Bromine Containing Environments*, 2nd World Congress on Desalination and Water Reuse, Volume II (Topsfield, MA: International Desalination Association, 1985).
12. ISO 9223, *Corrosion of Metals and Alloys-Corrosivity of Atmospheres-Classification, Determination and Estimation* (Geneva, Switzerland: International Standards Organization).
13. Fiegl, F., *Spot Tests in Inorganic Analysis* (Amsterdam, Holland: Elsevier Publishing, 1958).
14. Oldfield, J.W., and B. Todd, "Vapour Side Corrosion in MSF Plants," *Desalination* 66 (1987): pp. 171–184.
15. Page, G.G., "Influence of Syphonic Conditions on Erosion-Corrosion of Brass Heat Exchanger Tubes: A Review," *New Zealand Journal of Science* 26 (1983): p. 415.
16. Evans, G.J., and H.G. Masterton, "The Corrosion of Cupro-Nickel Feed Water Heater Tubes in CEGB Stations," *Proceedings of the 1st International Congress on Metallic Corrosion* (1961): p. 619.
17. Olsson, J.O., and H.L. Groth, "Evaporators Made of Solid Duplex Stainless Steel: A New Approach to Reduced Costs," International Desalination Conference (Topsfield, MA: International Desalination Association, 1992): p. 327.
18. Olsson, J.O., V. Jagerstrom, and I. Resini, "MSF Flash Chambers of Solid Duplex Stainless Steel," World Congress on Desalination and Water Reuse (Topsfield, MA: International Desalination Association, 2003).
19. Caddens, R., Weir Westgarth, Private Communication with R. Francis (2004).

7 Multiple Effect Distillation (MED)

Multiple effect distillation (MED) is a simple process where fresh water is condensed from vapor released from seawater. As described in Chapter 1, the process has been used for many years. The process has the advantage that it can come in a wide range of sizes; Figure 7.1 shows a multitrain MED plant attached to a power station.

FIGURE 7.1 A typical modern multitrain MED plant.

7.1 Method of Operation

MED plants, like MSF plants, require a source of heat, which can be steam or hot water or electricity, and so are often run in conjunction with a power station. In most MED plants, a thermo-compressor or a mechanical vapor compressor is added to improve efficiency. These are known as thermo-compression distillation (TCD) and mechanical vapor compression (MVC) plants, respectively. The higher efficiency of TCD plants means that they are the most common variant.

Figure 7.2 shows a schematic drawing of a typical 4-stage TCD plant, although it may contain from 2–14 stages. Vapor at ~ 75 °C is introduced into the first stage evaporator tubes, where it is condensed by externally sprayed raw water. In cooling the evaporator tubes, the raw water in stage 1 is heated and part of it vaporizes. The vapor enters the tubes of the second stage, is condensed by raw water, as in stage 1, forming more condensate. This process repeats itself in the subsequent stages. Part of the vapor produced in an intermediate stage (fourth stage in Figure 7.2) is drawn up by the thermo-compressor, which increases its pressure and mixes it with high pressure steam (Figure 7.2) to feed the first stage. The remainder of the vapor passes to the following stages and then to the final heat exchanger, where it condenses on the outside of tubes cooled internally by raw water. Part of this raw water becomes the feed to the evaporator stages and the rest goes to waste. The product water is collected and piped away (location 5 in Figure 7.2), while the brine goes to blow down (location 6).

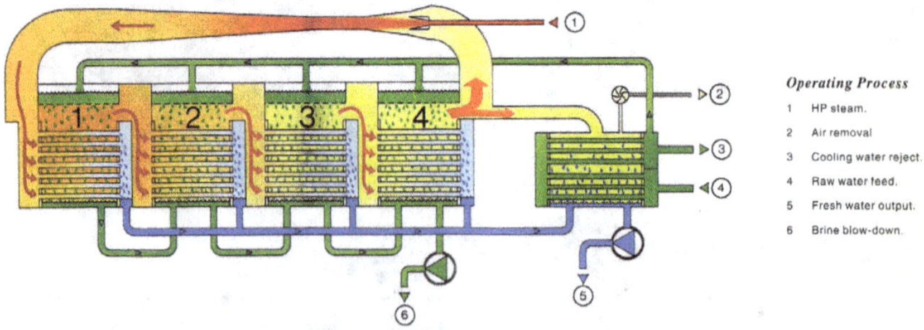

FIGURE 7.2 Schematic diagram of an MED plant.

An MED plant operates in the same way, but without the thermo-compressor. In an MVC plant, the thermo-compressor is replaced with a mechanical vapor compressor.

7.2 Conditions in MED Plants

The final condenser handles fully aerated seawater that may also be chlorinated to control fouling. Hence, the materials for this heat exchanger must resist this aggressive environment or be cathodically protected.

Although the raw seawater spray area is fully aerated, the dissolved oxygen content decreases farther down the evaporator, although it is not controlled, and does not drop as low as in MSF plants (< 20 ppb). This means that the brine collecting at the bottom of each evaporator stage is also low in oxygen. A number of chemicals may be added to the seawater prior to it going to the evaporators. If chlorine has been added to control fouling, then sodium metabisulfite, or a similar chemical, will be added to remove it. In addition, proprietary compounds will probably be added to control scaling and to prevent foaming. All of these are at low concentrations and they do not affect corrosion at these concentrations.

As in MSF plants, the incondensable gases (air and carbon dioxide) are vented, although not necessarily from each stage. In large plants with many stages, the incondensable gases may be cascaded down several stages to a vent system.

The product water is low in chloride and warm, and is usually treated before going into the drinking water system. The water treatment will vary depending on whether the water is going for boiler feed, for irrigation, or for drinking water. The most common use is for drinking, and materials for this are discussed below.

7.3 Component Materials

The commonly used materials for many of the components in an MED plant are discussed below. Because materials vary from plant to plant, the factors affecting the choice are discussed as well. In recent years, the relative cost of materials has changed and this is also discussed because it affects cost-effective materials selection.

7.3.1 Evaporators

The evaporator shells, baffles, tube sheets, and tube support plates were traditionally in 316L stainless steel. However, since 2000 the cost of duplex stainless steels compared to austenitic alloys has dropped, because duplex stainless steels contain less nickel and molybdenum. Currently, lean duplex 2304 is lower cost than 316L, and cylindrical evaporators are made of 2304, which has a similar corrosion resistance (see Chapter 4), but greater resistance to stress corrosion cracking (SCC) and higher strength, so section thicknesses can be reduced–within design code restraints–to further lower costs. Rectangular evaporators usually have a 2205 duplex floor and 2304 for the walls, roof, etc. These alloys are readily welded together using 2205 welding consumables.

The use of duplex stainless steels for rectangular evaporators means that not only are wall thicknesses reduced, but there are fewer stiffeners, further reducing costs. This means that an evaporator in duplex stainless steel is substantially cheaper than one in 316L.

The evaporator tubes were traditionally in aluminum brass except for the top three rows, which were titanium. The use of titanium here was to prevent erosion corrosion from the spray heads at the top of the evaporators. Erosion has been observed where three full rows of titanium were not fitted at the top. Because rows of heat exchanger tubes are offset, a full row is two offset rows so that water will encounter at least three titanium tubes before the aluminum brass.

However, the trend is now to use all titanium tubing in the evaporators, because the cost of titanium has dropped to little more than that of aluminum brass. Using titanium removes the risk of corrosion by the seawater. However, it is then necessary to provide extra support plates (compared with aluminum brass tubes) because titanium has reduced fatigue resistance compared with brass. Where it is a new design, the extra support plates can be included easily, but a re-tube in titanium from aluminum brass will mean the provision of supports between the tubes to reduce the risk of a fatigue failure. There is no risk of galvanic corrosion of the 2304 with titanium tubes in the evaporators because the water is treated to prevent the formation of a biofilm and it is also low in dissolved oxygen. Under these conditions, the potentials of the two alloys are very similar and titanium is a poor cathode under such conditions.[1]

7.3.2 Final Condenser

The final condenser is similar to the heat rejection section in an MSF plant in that water vapor is being condensed on the outside of the tubes, which have aerated seawater, possibly with chlorine additions to control fouling. Traditionally, the tubes were one of the copper-nickel alloys, but with the reduced price and their virtual immunity to the corrosion problems that can affect copper alloys (erosion corrosion, sulfide attack etc.[2]), titanium tubes have become more or less standard. To avoid fatigue failure problems it is necessary to provide additional tube support plates compared with copper alloy tubes, as discussed in the previous section.

The tube sheets used to be copper-nickel and then 316L stainless steel, with sacrificial mild steel anodes in the water boxes to prevent galvanic corrosion of the 316L. Nowadays, it is usual to use 2205 duplex stainless steel for the tube sheets, which can be of reduced thickness compared with 316L, because of the higher strength of 2205. Sacrificial mild steel anodes are still required with 2205 tube sheets, to prevent galvanic corrosion. The anodes should be mild steel ($C < 0.15\%$) to avoid fragmentation issues when they are corroding. They should be suitably sized to last one or two years before they need replacement. In addition, the water box should be suitably designed so that the incorporation of the anodes does not affect the water distribution, or introduce undue turbulence.

The water boxes used to be carbon steel lined with 90/10 copper-nickel or were rubber lined (e.g., neoprene), but more recently they were fabricated from 316L and the anode size was increased so that both the water box and tube sheet were protected. With the reduced cost of duplex stainless steels, the water boxes are now mostly made of 2205 duplex with protection from mild steel anodes. The higher strength of 2205 compared with 316L means that wall thicknesses can be reduced to lower costs.

On the vapor side the shell, baffles and tube support plates were commonly made of 316L stainless steel, although 304L was sometimes used for the baffles and tube support plates. With the reduced cost of duplex alloys (as discussed above), the shell, baffles, and tube support plates are made of 2304 duplex stainless steel or one of the other lean duplex alloys.

7.3.3 Venting System

In larger MED plants, typically more than 1,000 m^3/d, the vent system is a steam-ejector based system, usually in two stages, and the vents are cascaded down to the final condenser. There they are drawn into the first stage ejector and then to an intercondenser to remove most of the vapor before going to the second ejector, where the vapor is condensed in the after-condenser. Temperatures at the ejector discharge stages are high, being around 120–130 °C after the first stage, and around 140–160 °C for the second stage. This is similar to an MSF system, but without the precondenser—the function of which, in an MED plant, is served by the final condenser.

Materials are typically 316L as a minimum, but where corrosion is occurring, this could be upgraded to 904L, 6% Mo austenitic, or superduplex stainless steel, depending on price and availability. The condensers tend to be titanium–the same as the final condenser to avoid corrosion problems. Because these are relatively small units compared with the final condenser, there is no significant cost penalty. The other components in the condenser are usually the same as described above for the final condenser.

Where temperatures exceed 300 °C in the motive steam design, then duplex stainless steels are not suitable, because alpha prime phase will precipitate and greatly reduce ductility.[3] In this case, the materials must be austenitic stainless steel or titanium.

7.3.4 Pumps, Valves, and Piping

The main seawater intakes tend to be of a similar design to those used in MSF plants and are described in Chapter 6. For aerated seawater, fiberglass reinforced plastic (FRP) is the usual piping material, although small diameter piping is more commonly stainless steel. In aerated seawater, this should be superduplex or 6% Mo stainless steel to have good resistance to crevice corrosion, but lower alloy stainless steels have been used with varying degrees of success.

In low oxygen brine and produced water, the low oxygen or low chloride content means that lower alloy stainless steels are often satisfactory for smaller diameter piping, with 316L being the principal choice.

Piping for vent lines and chemical dosing is also usually in 316L stainless steel, and storage tanks for the chemicals are often austenitic stainless steel. However, 304L can suffer severe atmospheric corrosion and pitting in hot, humid climates because MED plants are usually adjacent to the sea, so chlorides are high. As a minimum, the tanks should be 316L in such environments, and in very aggressive atmospheres they may need to be 2205 duplex.[4]

Small valves (up to nominal pipe size, NPS, 2) are often ball valves and are made of 316L stainless steel. Larger valves are usually butterfly valves with a rubber lining on a steel or cast iron body. In aerated seawater, the discs should be NAB or superduplex stainless steel. If lower alloy stainless steels are used, crevice corrosion may occur under the seals if the valve is left closed for an extended period (days). In produced water (low chloride,) or brine (low oxygen), 316L discs are acceptable.

The main seawater pumps used to be austenitic cast iron (ACI) bodies with stainless steel impellers and shafts. However, the SCC problems with ACI, described in Chapter 4, mean that main seawater pumps should be in superduplex stainless steel to resist crevice corrosion. If lower alloy stainless steels are used, such as grade 1B or 4A (see Table A.4), there is a risk of crevice corrosion on the suction and discharge flanges. If 2205 duplex is used for the shaft, then crevice corrosion might occur at any of the locations where seals or impellers are located. Such a pit would be a stress raiser and would be followed shortly after by a costly fatigue failure. Shafts should *always* be superduplex stainless steel.

If a lower alloy stainless steel is used for the case, then crevice corrosion can be prevented on the flange connections by a high velocity oxy-fuel (HVOF) coating or a weld overlay of a more corrosion resistant nickel alloy. Alloy 625 is not suitable as a weld overlay because of the dendritic nature of the as-deposited alloy and its consequent low corrosion resistance (see Chapter 3), and it is necessary to use a nickel alloy such as C-22, C-2000, alloy 59, or alloy 686.[3, 5]

For deaerated brine and produced water, pumps in 316L, CD4MCuN cast duplex (see Table A.4), or cast 2205 will be acceptable. Chemical dosing pumps are relatively small and are usually 316L stainless steel, unless the chemical is aggressive as a concentrate.

7.3.5 Produced Water Treatment

The produced water may be treated in a number of ways depending on whether it is to be boiler feed water, drinking water, or irrigation water. Most of the time it is for drinking water and it needs to be further treated to make it suitable for distribution. First, it is necessary to stabilize the pH. This can be done by injecting CO_2 into the water and then passing the water through a limestone column to produce calcium bicarbonate, which will buffer the pH (as was described in Chapter 6 for MSF plants). However, an MED plant does not produce as much CO_2 as an MSF plant because it runs at a lower temperature. Hence, it is necessary to augment the CO_2 supply from a CO_2 generator. It is also necessary to add some mineral salts, so a little seawater may be added, or salty water from an SWRO plant, if there is one on site. Second, the pH is adjusted and chlorine is added to disinfect the water.

Because the water is low chloride it can be handled by 316L stainless steel, although the limestone column could be fabricated from 2304 lean duplex. Polymer piping may also be used provided the polymer is approved for potable water.

7.3.6 Instrument Tubing

Instrumentation operates in two environments, just the same as MSF plants. The first is aerated seawater and the most suitable materials are 316L and 90/10 copper-nickel, as discussed in detail in Section 5.3.5). It is important to select a suitable isolation valve and take note of the critical stand-up height to avoid corrosion problems.

The materials selection for instrumentation tubing in deaerated water is discussed in Chapter 6. The tubing options are the same (316L or 90/10 copper-nickel), but there is no critical stand-up height. There are considerations, however, that may result in one option or the other and these are discussed in detail in Section 6.3.9.

7.3.7 Ancillary Equipment

Materials for chemical storage tanks and chemical injection pipework are discussed in Chapter 5 for SWRO plants, and the same applies to such tanks and pipework in an MED plant.

There are many items of equipment that are not exposed to aggressive environments internally, but are exposed to the atmosphere. Such equipment includes electrical control and junction boxes, storage tanks, etc. These may be made of carbon steel or low-alloy stainless steel (304L or 316L). The atmosphere at many MED plants is both hot and laden with salt from their proximity to the sea. This means that it will be classed as C5 or CX, the most aggressive atmospheres, by ISO 9223.[6] Hence, carbon steel will require a high quality paint coating, and low-alloy stainless steel will require regular cleaning, and possibly painting. This is discussed in detail in Chapter 4 for carbon steel and stainless steel, respectively. The protection of FRP enclosures from UV is discussed in Chapter 2.

7.4 Service Experience

In recent years, the author has investigated two corrosion problems at MED plants. The first of these concerned corrosion of evaporator tubes.

An MED plant in the Middle East was suffering numerous failures in the evaporator tubes. The top three rows of tubes were titanium and the rest were aluminum brass. Figure 7.3 shows the appearance of some typical leaking tubes. The failures were mostly in the upper aluminum brass tubes and those down the sides of the evaporators. The appearance of the films on the tubes in Figure 7.3 is typical of those that form on aluminum brass in natural seawater.[2] The seawater was continuously chlorinated to control fouling and the concentration was measured three times per day at the settling tank. The seawater going to the spray nozzles was dosed with the usual antiscaling and antifoaming chemicals.

FIGURE 7.3 External appearance of aluminum brass evaporator tube.

At high magnification, the attack was clearly erosion corrosion, sometimes as a series of smaller pits and sometimes as one large pit (Figure 7.4). The polymer spray nozzles also showed clear signs of erosion corrosion (Figure 7.5).

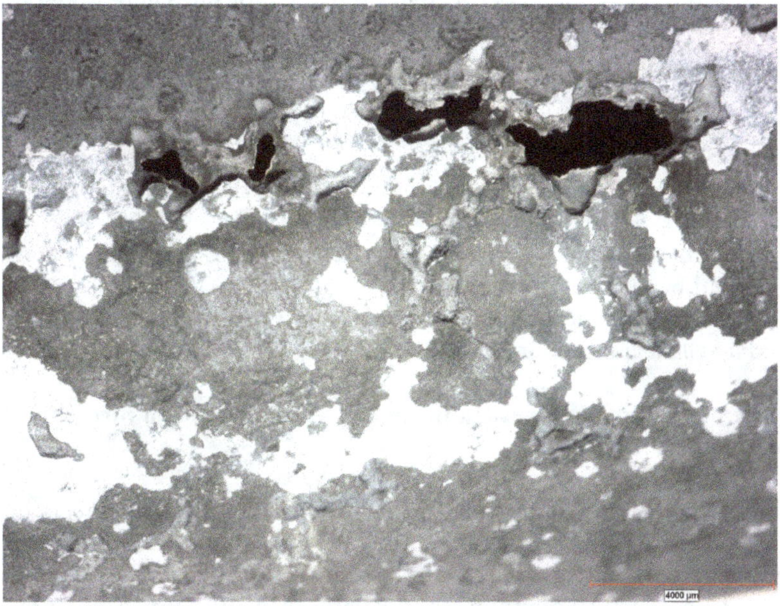

FIGURE 7.4 External erosion-corrosion of aluminum brass evaporator tube.

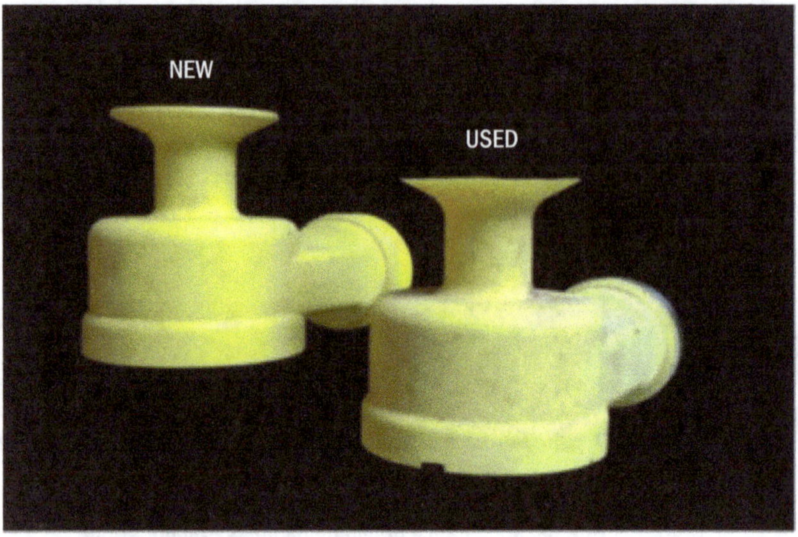

FIGURE 7.5 Appearance of new and used polymer spray heads.

The seawater intake was a 750 m closed culvert feeding water to a settling tank, which then fed all the MED trains. The settling tank was 4 m deep, but was found to contain between 0.9 and 1.5 m of silt. There were reports of rubble being dumped on the coast about 700 m from the main seawater intake as part of an extensive land reclamation project. Silt was clearly causing the erosion corrosion and it may have been washed down the coast from the reclamation project.

The following actions were suggested to prevent the erosion corrosion:

- Conducting a hydrology survey to see if settling of silt outside the plant could be encouraged;
- Dredging the settling tanks to clear all pump suction intakes; and
- Fit self-cleaning filters in each of the feed lines to the spray heads.

The combination of these actions should remove all risk of further erosion corrosion.

Another case concerned severe corrosion of the 316L tube sheets in the final condenser of an MED plant, which was fitted with titanium tubes. Alloys such as 316L, 2101, and 2304 will suffer accelerated corrosion in seawater when coupled to titanium without cathodic protection, as discussed in Chapter 4. Figure 7.6A shows the voluminous corrosion products between the tubes, while Figure 7.6B shows the deep pitting between the tubes after cleaning. This is normally prevented by inserting sacrificial, low-carbon steel anodes in each water box (as discussed in Section 7.3.2), but in this case, the anodes had accidentally been electrically isolated from the tube sheet and water box by extending the neoprene coating on the steel water box onto the threads of the mounting studs for the anodes. The problem was solved by high pressure water blasting of both tube sheets to remove the corrosion products, and filling of the pits with a cold-setting ceramic filler, in addition to removing the coating from the studs.

(A)

(B)

FIGURE 7.6 Appearance of 316L tube sheet with titanium tubes in the final heat exchanger in an MED plant: A) as opened, B) after cleaning, showing deep attack.

7.5 Future Trends

Even with the addition of antifoaming and antiscaling compounds, the temperature in an MED plant is limited to a maximum of around 75 °C. If cost-effective and reliable nanofiltration could be developed, then there would be no need for the addition of antiscaling compounds and MED plants could run hotter, thereby increasing the efficiency.

The likely maximum would then be around 110 °C, as currently used in MSF plants. This would not necessitate any major material changes because the ones currently in use could cope with this temperature. It might be necessary to upgrade some of the materials in the vent system to cope with the higher temperatures.

REFERENCES

1. Francis, R., *Galvanic Corrosion: A Practical Guide for Engineers*, 2nd ed. (Houston, TX: NACE International, 2017).
2. Francis, R., *The Corrosion of Copper and Its Alloys: A Practical Guide for Engineers* (Houston, TX: NACE International, 2010).
3. Francis, R., *The Corrosion of Duplex Stainless Steels: A Practical Guide for Engineers* (Houston, TX: NACE International, 2018).
4. Mameng, S.H., and L. Wegrelius, "Atmospheric Corrosion Resistance of Stainless Steel in the Middle East: Result of a Field Exposure Programme in Various Environments," CORROSION 2017, paper no. 8964 (Houston, TX: NACE International, 2017).
5. Rogne, T., J.M. Drugli, and T. Solem, "Crevice Corrosion Properties of Weld Overlays of Ni-Based Alloys Compared to 6%Mo Steels for Seawater Applications," CORROSION '98, paper no. 696 (Houston, TX: NACE International, 1998).
6. ISO 9223 (latest revision), "*Corrosion of Metals and Alloys-Corrosivity of Atmospheres-Classification, Determination and Estimation*" (Geneva, Switzerland: International Standards Organization).

8 Standards, Quality, and Procurement

In this chapter, the inadequacy of some material standards is demonstrated and guidance is given on how to technically specify materials to ensure that they will be fit for their intended purpose. The advisability of additional tests with some alloys is also discussed. The importance of quality assurance and quality control (QA/QC)—both with vendors and on delivery—is emphasized, as are the important factors affecting satisfactory fabrication.

8.1 STANDARDS AND SPECIFICATIONS

8.1.1 Metals

When procuring metals and alloys it is common to use ASTM specifications (e.g., ASTM A240 for stainless steel plate and sheet). However, these are product standards that are written largely by material producers to achieve minimum acceptable requirements and tend to focus on the composition and mechanical properties. Although these standards may make a brief mention about suitable heat treatment, they do not go into detail about how to achieve the best properties, or how to test to ensure the product has been properly manufactured. With the worldwide increase in the number of corrosion resistant alloy (CRA) vendors, and increasing project desire to minimize costs, it is essential that additional testing, over and above that required by ASTM, is written into a supplementary specification to ensure the supply of good quality material. This principal has been adopted by the Norwegian standards organization, Standards Norway (NORSOK). These standards cover a range of materials, and many companies use NORSOK M-630,[1] or incorporate its requirements into their company specifications.

There is also a new standard, ISO 17781, which provides details of the testing requirements to ensure good-quality duplex stainless steel is procured.[2] Many feel this standard will be the long-term successor to NORSOK M-630.

8.1.2 Polymers

Advances in polymer chemistry, and the understanding of how to blend and formulate polymers, allow for very specific grades of a particular polymeric material to be produced. It is almost never sufficient to specify polymeric material generically, and the grade of polymer must be matched to the end-use application. For example, polypropylene is commonly used for handling caustic solutions, but the block copolymer (grade H2222) offers better chemical degradation and cracking resistance than the homopolymer. In general, for many chemical applications, unfilled polymers, without plasticizers, have the optimum chemical resistance. If other grades are used, then exposure testing of the intended material in the target environment will provide confirmation that the material is acceptable.

For fiberglass reinforced plastic (FRP) for seawater use, the epoxy resin is usually adequate. However, it is important to specify a UV-resistant outer coating and additional measures, for whatever resin is specified, because FRP pipe is mostly exposed to strong sunlight in desalination plants, as discussed in Chapter 2.

8.2 TEST REQUIREMENTS

Below are some suggestions for testing required to ensure the supply of good quality CRAs, based on the published experiences of both the author and others.

8.2.1 Austenitic Cast Iron

There are a number of grades of austenitic cast iron (ACI), as discussed in Chapter 2. The alloys D2 or D2B are the ones preferred for good resistance to corrosion in seawater, while retaining some weldability. All ACI alloys will suffer stress corrosion cracking (SCC) in warm seawater because the stresses at the interfaces between different particles in the structure are sufficient to initiate attack. The initiation of SCC is delayed by ensuring all castings are fully stress relieved.

Thickening the castings also reduces stresses and delays failure, but with the high nickel content, this introduces a cost penalty.

8.2.2 Copper Alloys

8.2.2.1 Brasses

In desalination plants, the principal brass alloy is aluminum brass; it is used as heat exchanger tubes in older multiple-effect distillation (MED) plants, and in the lower temperature heat recovery section in multi-stage flash distillation (MSF) plants. In addition to the usual tests for dimensional tolerances, straightness, ammonia cracking resistance, etc., it is important to test for residual carbon films from the manufacturing process. These are well known to cause pitting of copper tubes in certain fresh waters, but the author has also seen failures of copper alloy heat exchanger tubes in seawater due to carbon films.[3]

Many standards (such as ASTM B359 for seamless copper alloy heat exchanger tubes) do not make any mention of carbon films or other deleterious films in the bore. The current European standard for copper alloy heat exchanger tubes, EN 12451, does not specifically require that heat exchanger tubes should be free from films in the bore, but leaves it up to the purchaser.[4] It is important to specify that samples of tube will be tested using a thermal test, such as in the LECO-type test.[(1)] In this test, the sample is heated in a stream of oxygen and the CO_2 produced at different temperatures is determined. By heating at increasing temperatures, the carbon originating from various sources can be isolated:

- solvents at 200 °C
- greases at 650 °C
- carbon films at 850–900 °C

Only the carbon measured at the highest temperature is used for assessing the tube quality. The pass/fail criterion that has been adopted is the same as that used for copper tubes for use in fresh waters, i.e., carbon ≤ 0.2 mg/dm^2. Tubes that fail this test can be cleaned by internal grit blasting or by pickling, and then retested.

Naval brass has been used with copper alloy tubes for tube sheets, and although it dezincifies, the tube sheet is thick (≥ 35 mm) and the rate of propagation is low, so a failure is unlikely.

1. The LECO-type test is a registered trademark of LECO Corporation, St. Joseph, MI.

Manganese bronze is actually a high tensile brass and this will dezincify rapidly in seawater. If a high strength copper alloy is required, then it should be nickel aluminum bronze (NAB), although nickel alloy K-500 and duplex stainless steel are also used where high strength is required. The additional tests for these alloys are discussed below.

8.2.2.2 Copper-Nickel Alloys

Copper-nickel alloys being used for heat exchanger tubes should be subject to the same tests as aluminum brass, as discussed in the previous section, including the LECO test for carbon films, with the same pass/fail criteria.

The 90/10 copper-nickel being used as plate, sheet, or pipe may also suffer from another problem. The alloy gives the best corrosion resistance when the iron content is in the range 1.5 to –2.0 wt%.[5] However, all the common standards for the alloy give the iron concentration as 1.0 to –2.0 wt%. To ensure adequate resistance to erosion-corrosion it is necessary to specify that the minimum iron content must be 1.5 wt%.

This level of iron exceeds the solubility limit for the alloy at room temperature. However, if the iron is not in solution, the erosion-corrosion resistance decreases quite markedly.[5] To retain the iron in solution, it is necessary to rapidly cool the alloy from an annealing temperature of ~ 900 °C. It has been stated that this problem is well understood and modern manufacturing methods will always produce good quality material. However, the author examined failures of 90/10 copper-nickel pipes from the cooling water system of the Saltire platform in the North Sea. The pipes were suffering a very high rate of general corrosion plus erosion-corrosion after the bends, even though the water velocity was only 2–3 m/s. The clue to the cause of the problem was at the welds. Not only was the 70/30 copper-nickel weld metal not attacked, neither was the parent metal in the heat-affected zone (HAZ). Figure 8.1 shows a half-section of pipe; the cut edge clearly shows the thicker metal in the weld and HAZ. This was because the iron was not in solution in the pipe, but the weld thermal cycles in the HAZ had resolutionized the iron and restored the corrosion resistance. The weld metal was not affected because iron is more soluble in the 70/30 copper-nickel filler metal. Determining the magnetic permeability with a simple device, such as the Severn magnetic gauge (Severn Engineering Co. Inc., Auburn, AL), showed the parent metal to have a value of > 1.2, while the weld metal and HAZ were < 1.05. It is recommended that this simple check should be carried out on the parent metal at the same intervals as the other tests (e.g., for carbon films and residual stresses). Material with a magnetic permeability of > 1.10 should be rejected.

FIGURE 8.1 Selective attack of parent 90/10 copper-nickel due to most of the iron being out of solution.

8.2.2.3 Nickel Aluminum Bronze

NAB is a widely used alloy in both cast and wrought forms. NAB has sometimes been observed to suffer dealuminification. It has been found that keeping the nickel content ≥ 0.5% above the iron content will prevent dealloying.[6] This is not covered by most specifications and should be added as a requirement when specifying this alloy, in either cast or wrought form.

At the high temperatures used in hot working or during welding, NAB has a structure that is alpha and beta. During cooling, the beta phase transforms to alpha and kappa. However, if the cooling rate is too fast, the beta phase does not fully transform.[6] The beta phase is highly anodic to the alpha phase and it is attacked rapidly in seawater. The (UK) Royal Navy experienced corrosion problems with weld-repaired castings in NAB, while seam-welded pipe in a desalination plant corroded and "unzipped" along the weld (see Chapter 3), both due to residual beta phase.

The author was involved in investigating corrosion at a large MSF desalination plant in the Middle East, where 2 NAB tube plates out of 128 showed pitting due to the presence of beta phase. It was thought that these two were the first in a batch to be rolled, and cooled too quickly off the rolls. As subsequent plates piled up, the thermal mass increased and cooling rates slowed. The tube plates could not be weld repaired because that would have created more beta phase. The solution was to grind out the corrosion and fill with a cold-setting ceramic filler.

It is recommended that all NAB be given a heat treatment after the final hot operation (welding, rolling, etc.). For castings, it is recommended that the Royal Navy heat treatment of 675 ° ± 25 °C for 2–6 hours, followed by air cooling is specified. For wrought products (such as thick plate) a heat treatment of 725 ° ± 25 °C for the same period followed by air cooling is recommended.[3] The author has specified this heat treatment for the NAB tube plates in another MSF plant and there have been no problems after more than 20 years in service.

8.2.3 Stainless Steels

8.2.3.1 Austenitic Stainless Steels

It must be accepted that all 300-series austenitic stainless steels, such as 316L and 304L, will be made to the minimum composition in the standard. For example, the requirement is that 316L contains between 2.0 and 3.0 wt% molybdenum and virtually all producers make 316L with only slightly more than the minimum 2.0%. Although standards exist for 316L with a higher minimum molybdenum content, this variant is not readily available unless a mill-run quantity is required. It is sometimes worth requiring an ASTM A262 Method E[7] test for 300-series alloys to check that the alloy is L grade and not a higher carbon version that, when welded, will form carbides at grain boundaries. The carbides remove chromium from the matrix and reduce corrosion resistance.

The higher alloy austenitic stainless steels, such as the 6% Mo grades, should have the additional tests required by NORSOK standard M-630.[1] These tests are conducted on a microsection (suitably etched and viewed at x500), to prove that the structure is free of sigma phase; and an ASTM G48 Method A corrosion test is conducted at 50 °C for 24 hours to show that there are insufficient deleterious phases to significantly affect the corrosion resistance.[8] The "pass" criteria for the G48A test are (1) no pitting, and (2) a weight loss less than 4 g/m^2. These criteria ensure that there are no significant deleterious phases or chromium-denuded zones, due to poor pickling. It is important to test samples with the same finish that will go into service.

8.2.3.2 Duplex Stainless Steels

There have been some well-publicized problems with duplex and super-duplex stainless steels due to poor manufacturing procedures.[9-12] As an example, Figure 8.2 shows the microstructure of a superduplex stainless steel fitting containing sigma phase due to poor heat treatment. The fitting failed after a few months in service in seawater. Francis et al. argued that the requirements of ASTM A923[13] for duplex stainless steel quality are inadequate and that even NORSOK requirements need strengthening.[14-15] This has been discussed further by Byrne, who also shows some failures due to poor quality manufacturing.[15] They suggested the following tests for 25% Cr superduplex alloys:

1. Impact toughness ≥ 70J average at −50 °C.
2. ASTM G48A test for 24 hours at 50 °C (no pitting and weight loss < 4 g/m^2).
3. Microsection at x500 (electrolytically etched in oxalic acid, then KOH or NaOH). No third phases to be present. Ferrite to be 40–60%

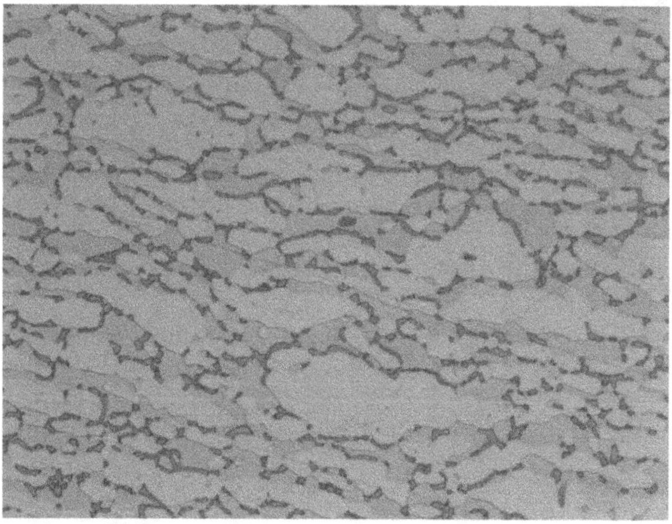

FIGURE 8.2 Microstructure of thin-walled duplex stainless steel fitting with sigma phase.

The requirements for 22% Cr duplex are the same, except the ASTM G48A test temperature is 25 °C.

The impact toughness test guarantees that, even after welding, there will be adequate toughness for most applications. Because third phases also reduce toughness, it is also a measure of the quality of the microstructure. Passing the corrosion test shows that deleterious third phases are not present in sufficient amounts to significantly affect the corrosion resistance. The corrosion test can also detect nitrides, which may cause pitting or just a high weight loss.[16] The weight loss criterion is also useful because it can detect surface chromium depletion due to inadequate pickling after heat treatment and quenching, even though this does not cause pitting. If the chromium denudation is very shallow, pitting does not occur because of the sound metal beneath the denuded zone. A high weight loss may also be due to inadequate scale removal during pickling. As is the case with high-alloy austenitic alloys, it is essential that the finish to be used in service is tested.

Using the microsection enables both the phase balance to be determined and to see any third phases, if the sample is polished and etched correctly.[15] This is particularly helpful if the material has failed test 1 or 2, because it helps diagnose the nature of the problem, identify possible root causes, and establish remedial action. If a duplex alloy contains too many third phase precipitates, or precipitates above a certain size, it would be impossible to pass either the impact test or the corrosion test.

The efficacy of these tests is demonstrated by some case histories in Reference 15.

Tests to detect deleterious phases in lean duplex alloys are covered in ASTM A1084, and are similar to those for 2205 and superduplex discussed above.[17] The standard requires two things:

1. Impact toughness ≥ 70J (2101) or ≥ 100J (2304) average at room temperature.
2. Modified ferric chloride test for 24 hours at 25 °C (no pitting and weight loss ≤ 10 mdd).

A microsection examination and ferrite count are not included in this test, but it is strongly recommended that it be included for lean duplex alloys, for the same reason that it is needed for 2205 and superduplex.

The higher temperature for the impact toughness test is because lean duplex alloys are low in nickel and have reduced low-temperature toughness compared with 2205 and superduplex, although it is acceptable for many applications.

Because lean duplex alloys have a lower PREN than 2205, they will suffer pitting in the ASTM G48A test at room temperature. Hence, a less aggressive test is specified, using a lower ferric chloride concentration plus an addition of sodium nitrate to act as a partial inhibitor. It is important that whatever finish is on the final product in service is what should be exposed in the corrosion test.

8.2.4 Nickel Alloys

The specifications for many older nickel alloys permit relatively high carbon levels, because they were written before the widespread introduction of argon oxygen decarburization (AOD) refining. Nickel alloys with higher carbon contents may have grain boundary precipitates that can lower corrosion resistance and lead to intergranular corrosion. Carrying out an ASTM G28A test will determine whether the alloy has been suitably manufactured and heat treated.[18] The pass-fail criteria need to be determined by comparing the corrosion rate for properly annealed material, as described in the references in ASTM G28.

Table 8.1 shows the corrosion rates for some common nickel alloys sufficient to just show intergranular corrosion in the G28A test.[19] However, from a commercial point of view, this is just the threshold and a common fail in this test is a corrosion rate > 1 mm/y, except for alloy C-276 (UNS N10276), which is often > 12 mm/y. Note that these rates are for wrought material. Welds and cast products can have a higher corrosion rate in a satisfactory condition and pass/fail criteria must be established for these separately.

TABLE 8.1 Corrosion rates of some common nickel alloys sufficient to just show intergranular corrosion in the ASTM G28 Method A test.[19]

ALLOY	UNS No.	TEST DURATION (hr)	CORROSION RATE (mm/y)
20	N08020	120	0.61
825	N08825	120	0.46
625	N06625	120	0.46
C-276	N10276	24	9.14

Nickel-based alloys are available in a number of heat treatment conditions. Make sure that the heat treatment condition being ordered has the mechanical properties for the application. This sounds elementary, but the author was involved in a project where this was not done and he was just in time to stop a large order for a nickel alloy being placed in the wrong heat treatment condition.

Alloy K-500 may be supplied in the annealed or the aged condition. If components are to be machined, then it is easier in the annealed condition and aging must be carried out subsequently to obtain the required strength. Such requirements need to be included in the technical specification.

8.2.5 Titanium Alloys

The properties of titanium alloys are very susceptible to a reduction caused by small changes in the content of oxygen, hydrogen, and particularly, iron. Be prepared to check these independently if using a new supplier. High residuals can result in an alloy that is unweldable or has reduced corrosion resistance. Where cyclic stresses are involved extra-low interstitial (ELI) material is preferred to give better fatigue resistance, and it is important to check that this has been specified and delivered.

8.3 IMPORTANCE OF QA/QC

8.3.1 Metals and Alloys

Having specified the above tests, care must be taken, because the certificates may not be true. The author has seen large diameter superduplex bar (420 mm) claiming to meet ASTM properties. When the small print is read on the certificate, or the question is asked, it turns out the test samples were taken from near-surface, while ASTM requires them to be at mid-radius (¼T, where T is the diameter). Samples from deeper below the surface fail because of the presence of third phases. It is not currently possible to quench duplex stainless steel thicker than 250 mm to avoid the formation of sigma phase.[20] This example emphasizes both the need for witnessing the testing, unless the supplier has been fully audited, and also the importance of QA/QC monitoring throughout. This costs money, but it costs significantly less than that of a major failure. Customers want to lower costs, but producers with limited experience may cut corners without realizing the implications.

Many companies rely heavily on NORSOK standards, M-630 for materials and M-650 for vendor approvals.[1,21] NORSOK M-650 was developed as a way of prequalifying manufacturers of "difficult" materials (largely duplex and superduplex) to ensure previous costly delays in production of components in these alloys were avoided. ISO 17782 is currently in development and this is also a vendor approval document.[22] However, just because a vendor claims to be NORSOK approved, be wary. Check when they were last audited–NORSOK certification lasts five years, but auditing should be done every three or four years, because manufacturers often change things in the production cycle or install new equipment that has not been audited. Auditing can be expensive, but it usually pays to do it when items are needed in a hurry, such as for urgent repairs.

In addition, check what materials the company is approved to supply–including both product forms and thickness range–*and only purchase what has been approved.* There is usually a good reason if they do not have approval over the full thickness range. It is essential to ensure that any potential supplier is approved for the specific material requirements needed.

Another document that is not as widely used as it should be is EEMUA 218, which gives details on how to obtain good quality duplex stainless steels.[23] Without going into details, its main points are the following:

- NORSOK M-650 preferred, but good historic data acceptable
- Audits recommended more frequently than every five years
- UKAS[(2)] accredited test laboratories only
- All testing at the center (½T)
- All parts of the manufacturing route and heat treatment must be covered in detail
- Surveillance and witnessing during manufacture
- A technical compliance audit required, particularly for new suppliers—*over and above a quality assurance audit*
- Stockists to be treated in the same way as manufacturers, with the same responsibilities (material bought from inventory to be subject to independent third party testing to verify the mill certificate properties)
- Positive material identification (PMI) and feritscope checks required

In addition to procurement specifications and monitoring during manufacture, it is worthwhile checking material on delivery with some simple nondestructive testing. Hardness tests can sometimes be useful, and Francis et al. describe the use of the feritscope to assess the quality of duplex stainless steels.[24] They comment on the factors that affect the readings, and show that significant amounts of sigma/chi phase produce a significant reduction in the feritscope reading. It is possible to take a lot of readings very quickly in many locations, and looking at the range of the readings suggests when there could be precipitates.

2. UKAS is the United Kingdom's National Accreditation Body

- Readings > 40 are OK
- Readings in the 30s could show problems if inconsistent, and further investigation is required
- Readings < 30 are a strong indication of sigma phase

Suspect material should then be checked with nondestructive or destructive metallography, because there are causes other than sigma that can give low feritscope readings, such as poor pickling.[24]

The problem fittings in Reference 9 passed through five companies from the manufacturer to the end user and not one of them ran a feritscope check, which would quickly have identified the suspect fittings. Francis et al. describe a case study where hardness checks on a bar found hard spots, and the feritscope confirmed that sigma was present in these areas.[24] This was caused by the use of unsuitable supports in the annealing furnace for the bar, which produced cold spots during heat treatment.

In addition to this, there are simple nondestructive tests that can be applied on delivery to try and identify problems early on. One of these is the use of random PMI checks, using a portable X-ray analyzer, to check the composition of all CRAs. Although these instruments are expensive (but the costs are coming down), only one batch of rogue material has to be identified to make up the cost of the analyzer. The author has been involved in some failure analyses where the supposedly high-alloy stainless steel supplied was, in fact, 316L. PMI quickly identifies errors of this kind on delivery. Alloy mix up is a significant issue in the mills, stockholders, distributors, and fabricators. Color coding at the mill is a good way of avoiding alloy mix ups.

With nickel alloys, and particularly heat treatable, aged alloys, a simple hardness test can confirm that the alloy is in the correct heat treatment condition. If it is not in the correct condition, this could lead to the further aging of aged material, which could significantly reduce ductility, or it could lead to the nonaging of solution annealed material, which would have too low mechanical strength.

The use of a simple gauge to measure relative magnetic permeability of 90/10 copper-nickel was mentioned above, and this can be used as a quick check on delivery.

With many of these requirements, it is often simpler to have QA/QC staff, or a third party witness, on site to verify that everything has been done correctly to the technical specification. Although this adds to costs, it is far less than that of a major failure, such as that described in Reference 9, where the cost was estimated to be several million dollars.

8.3.2 Polymers

Polymer formulations may differ significantly between manufacturers, so for demanding applications, polymers are often specified by the manufacturer's trade name and specific grade. Acceptance tests can include Fourier transformation infrared spectroscopy (FTIR) or differential scanning calorimetry (DSC) done on samples of batches of material. Specific gravity tests can also be useful to differentiate between polymers.

8.4 CASTINGS

The above discussion has principally addressed wrought products, but poor foundry practice can lead to solidification cracks, cold cracks, high levels of segregation, shrinkage, porosity, local high stresses, or an undesirable microstructure, all of which may lead to premature failure. When selecting a foundry for CRA castings, particularly larger ones, there are some factors to take into account:

- What is their experience? Many CRAs require some skill to cast correctly and it is always best to use a foundry with prior experience.
- Do they have computer modeling for solidification and stress distribution? The design of running, gating, and positioning of feeders or padding can affect the porosity, residual stress, and tear formation in castings. The ability to model this before pouring any metal can save a lot of time and money.[25]
- Is the heat treatment furnace adequate? The importance of this is discussed in Reference 20.
- Is the quench tank satisfactory? See Reference 20.
- Can they do fast transfers? See Reference 20.

Poor heat treatment can also cause problems with duplex stainless steel castings.[25] The additional tests (impact toughness, ASTM G48, and a microsection) described above for wrought products should also be applied to castings. ASTM A995 grades 1B and 4A should use the 25 °C temperature for the corrosion test, while grades 5A and 6A should use the superduplex temperature of 50 °C. Suitable tests for cast lean-duplex alloys have yet to be determined, but the new ISO 17781 gives some guidance.[2]

When procuring duplex castings, it is usual to start with ASTM A995 and add the extra test requirements under the testing options.[26]

The author has looked at numerous failures of duplex castings caused by poor quality due to a lack of adequate specifications and a lack of suitable QA/QC.

Castings in NAB should also have a heat treatment as the final stage after any weld repair and before machining, to optimize both the strength and the corrosion resistance. The recommended heat treatment is 675 ± 25 °C for 2–6 hours (depending on thickness), followed by air cooling.[6]

With all castings, especially large ones, it is recommended that a map be required showing all the sites where weld repair has been carried out and what postweld heat treatment (PWHT) was carried out, preferably with a third party witness. The importance of PWHT for NAB was discussed above, and ASTM A995 requires solution annealing after major weld repair on all duplex stainless steel castings.

8.5 FABRICATION

Having purchased the correct alloys, they will usually require fabrication. Ideally, a fabricator with a proven track record in welding the alloy(s) of concern should be used, because some CRAs require a level of skill that needs time to learn, e.g., titanium. Fabrication in the shop is preferred to site welding for CRAs, because it is much easier to control the welding conditions and cleanliness. Where sufficiently skilled site welders are in short supply, it has proved effective with some seawater reverse osmosis (SWRO) plants to spool the piping in a shop, and ship to site, where most of the assembly is by mechanical joining.

When welding many CRAs, there will inevitably be some heat tint in the HAZ. With stainless steels and nickel alloys, the thin layer of oxide has taken chromium from the matrix, which now has a reduced corrosion resistance under the heat tint. The heat tint should be removed either by abrasion, such as a stainless steel wire brush, or a suitable pickling paste. Mechanical grinding should be avoided, because it can introduce microcrevices into the metal surface, which are preferential sites for the initiation of crevice corrosion.[27] If iron or steel debris is left as contamination after cleaning, it can give rise to corrosion of lower alloy stainless steels (300 series) and it can produce unsightly staining on high alloyed grades, after coming in contact with moisture (see Section 8.5.3).

It is important to remove the heat tint on the side exposed to the process fluid where possible, particularly for 300 series stainless steels. These grades are the most susceptible to corrosion under the heat tint because of their lower alloy content. If piping is being spooled, it can be feasible to get the spools pickled at a cleaning plant with large pickling baths. The author has seen this done on some SWRO projects.

Chapter 3 discussed the welding of each alloy type, but for the higher alloy CRAs, there are some tests that should be carried out in addition to those specified in ASME IX, or whatever fabrication standard is being followed. These tests should be carried out for every Procedure Qualification Record (PQR) weld (and weld repair PQR for castings), as well as every weld made for individual welder qualifications.

8.5.1 Austenitic Cast Iron

The problems with welding ACI are discussed in Chapter 3, but if welding is carried out, it is important to ensure that all welds are stress relieved at 450–550 °C, to minimize the risk of SCC in warm seawater.

8.5.2 Copper Alloys

The importance of selecting the correct filler when welding copper alloys was discussed in Chapter 3. When welding NAB with matching filler, it is essential to carry out a PWHT to ensure that all residual beta phase is transformed to alpha and kappa. Beta phase is highly cathodic to alpha phase in seawater and will corrode preferentially and rapidly (see Chapters 3 and 4). The recommended heat treatment is that specified by the Royal Navy, 675 ° ± 25 °C for 2–6 hours, followed by air cooling.

8.5.3 Stainless Steels

8.5.3.1 Austenitic Stainless Steels

The selection of fillers for low-alloy austenitic stainless steels was discussed in Chapter 3, and the importance of removing the heat tint was discussed above. Where mechanical cleaning is carried out, it is important to use equipment dedicated to stainless steels, because contamination of 300 series alloys with carbon steel can cause corrosion problems. The corrosion of the carbon steel not only causes unsightly rust stains, the pH of the solution below the rust can be low enough to initiate pitting of the stainless steel.

With high-alloy austenitic stainless steels, a microsection should be prepared, to show (at x500) that there are no deleterious phases, such as sigma, present or an undesirable unmixed zone.[28] In addition, an ASTM G48A corrosion test should be carried out at 40 °C for 24 hours. The pass criteria are no pitting and a weight loss < 4 g/m². This demonstrates that the weld has not got any deleterious phases or depleted zones that significantly affect the corrosion resistance. These requirements are included in NORSOK M-601.[29]

8.5.3.2 Duplex Stainless Steels

In addition to the requirements of ASME IX or whatever welding code is being used, the following tests should be carried out. The tests are similar to those in Section 8.2.3.2 for parent metal, but with slightly different test conditions or pass criteria. For superduplex stainless steel they are listed below:

1. Impact toughness ≥ 40J average, 35J minimum single value, at −50 °C, in the weld metal, fusion line, fusion line + 2 mm and fusion line + 5 mm.
2. ASTM G48A test for 24 hours at 35 °C (no pitting and weight loss < 4 g/m²).
3. Microsection at x500 (electrolytically etched in oxalic acid, then KOH or NaOH). Ferrite to be 35–65%.

The requirements for 22% Cr duplex are the same, except the ASTM G48A test temperature is 20 °C. The justification for these tests is exactly the same as that in Section 8.2.3.2 for parent metal. Such testing is a common requirement in the oil and gas, and chemical and process industries. All corrosion test pieces should have the weld root exposed in the as-welded condition, unless the welds will be machined prior to entering service.

Surface contamination on stainless alloys can have a marked influence on corrosion resistance and its removal must be considered whenever equipment is being fabricated. If there is a need to have heat tint removed, the costs for heat tint removal and postweld clean up must be included in cost estimates. Figure 8.3 shows rusting of iron contamination on superduplex stainless steel welded spools, due to inadequate cleaning.

FIGURE 8.3 Rusting due to iron contamination from poor cleaning of superduplex stainless steel welds. [Photo courtesy of Rolled Alloys]

At the time of writing, tests for lean duplex stainless steel welds have not been specifically addressed. The author feels that the impact toughness test as shown above should be suitable, but carried out at room temperature, with the same pass/fail criteria. The modified ferric chloride test (as described in Section 8.2.3.2) would be useful, but the temperature would depend on the filler. A test at 20 °C might be suitable, but validation tests would be required for verification. A microsection should also be carried out, as for 2205 and superduplex, for the same reasons. All corrosion test pieces should have the weld root exposed in the as-welded condition, unless the welds will be machined prior to entering service.

8.5.4 Nickel Alloys

Many of the older nickel-based alloys, such as alloy 20, have a rather high maximum carbon content by modern standards, e.g., for alloy 20 it is 0.07%. Depending on the carbon content of the parent metal and weld filler, there may be carbides in the structure, making the alloy susceptible to intergranular corrosion. It is suggested that all nickel alloy welds have an ASTM G28 Method A test specified, unless both parent and filler have a carbon content ≤ 0.03 wt%. The pass/fail criteria suggested in Section 8.2.4 for wrought products are not generally applicable to welds and it will be necessary to compare the microstructure of some welds with the corrosion rate to establish a suitable pass/fail criterion.

NOSOK standard M-601 recommends the same tests for Ni-Cr-Mo alloys as specified for 6% Mo austenitic stainless steels.[29] That is a suitably etched microsection at x500, to show the absence of deleterious phases, and an ASTM G48A test at 40 °C for 24 hours. There should be no pitting and a weight loss < 4 g/m^2. The samples should have the same surface finish as will be used in service.

8.5.5 Titanium Alloys

Titanium alloys should ideally be free from weld discoloration. Welders must ensure not only that the actual weld pool is shielded, but that the hot weld is maintained in an inert atmosphere until it has cooled to well below 400 °C. Oxidation above 450 °C results in a transformation to alpha case at the surface, and as a result, surface ductility, fatigue strength, and corrosion resistance are reduced. If such a layer is found, it must be removed. The reader is directed to an excellent summary of the issue and a color comparator by Talkington et al.[30]

8.5.6 Polymers

Joints in FRP piping systems are generally glued for tapered joints up to 150 mm diameter pipe, and butt and strap joints are used for larger diameters. Figure 8.4 shows some typical tapered joints awaiting assembly. In some applications, butt and strap joints may be used for smaller diameter pipes. It is critical that the workers who are employed to make these joints are trained and experienced. Typically, the pipe manufacturer will supply joint kits, which consist of precut veil and reinforcement that can be applied on site by suitably trained and competent personnel. In remote locations, it is common for the pipe vendor to send a supervisor to the site. The supervisor trains the local workforce to perform the joining work.

With FRP pipe, flanged joints are required to connect to other equipment, such as valves, and the flanges are usually attached to the pipes by butt and strap connections. Flange-to-pipe connections are typically made in the pipe shop and final, on-site joints are in the pipe runs.

FIGURE 8.4 FRP pipe with tapers for joining.

REFERENCES

1. NORSOK M-630, "Material Data Sheets for Piping, Edition 5" (Oslo, Norway: Standards Norway, 2010).
2. ISO 17781 (latest revision), "Petroleum, Petrochemical and Natural Gas Industries–Test Methods for Quality Control of Microstructure of Ferritic/Austenitic (Duplex) Stainless Steels" (Geneva, Switzerland: International Standards Organization).
3. Francis, R., *The Corrosion of Copper and Its Alloys: A Practical Guide for Engineers* (Houston, TX: NACE International, 2010).
4. EN 12451:2012, "Copper and copper alloys. Seamless, round tubes for heat exchangers," (Brussels, Belgium: Comite Europeen de Normalisation [European Committee for Standardization], 2012).
5. Bailey, G.L., "Copper-Nickel-Iron Alloys Resistance to Seawater Corrosion," *Journal of the Institute of Metals* 79, 5 (1951): pp. 243–292.
6. Campbell, H.S., "Aluminium Bronze Alloys Corrosion Resistance Guide," Publication No. 80 (Hemmel Hempstead, United Kingdom: (UK) Copper Development Association, 1981).
7. ASTM A262 (latest revision), "Standard Practices for Detecting Susceptibility to Intergranular Attack in Austenitic Stainless Steels" (West Conshohocken, PA: ASTM International).

8. ASTM G48 (latest revision), "Standard Test Methods for Pitting and Crevice Corrosion Resistance of Stainless Steels and Related Alloys by Use of Ferric Chloride Solution" (West Conshohocken, PA: ASTM International).
9. Rorvik, G., L.A. Marken, I.M. Kulbotte, M. Aursand, S. Olsen, and K.S. Karlsen, "Influence of Intermetallic Precipitates on the Mechanical Properties and Environmental Cracking Resistance of Duplex Stainless Steel Fittings-A Case History (Part 1)," Duplex Stainless Steels 2010 (Zutphen, Netherlands: KCI Publishing, 2010).
10. Mahajanam, S., H. Rincon, and D. McIntyre, "Metallurgical Defects in Duplex Stainless Steel Fittings," *Materials Performance* 49, 4 (2010): p. 56.
11. Ryengen, E., and C. Wintermark, "Lessons Learned from Heat Treatment of Components in 22Cr and 25Cr Duplex Stainless Steel (and other materials)," Duplex Stainless Steels 2010 (Zutphen, Netherlands: KCI Publishing, 2010).
12. Turbeville, E., F. Busschaert, S. Benum, and B. Hadsen, "A Summary of Recent Experiences on Designing and Fabricating Subsea Oil and Gas Production Systems with 22Cr and 25Cr Piping Materials," Duplex Stainless Steels 2010 (Zutphen, Netherlands: KCI Publishing, 2010).
13. ASTM A923 (latest revision), "Standard Test Methods for Detecting Detrimental Intermetallics in Duplex Austenitic/Ferritic Stainless Steels" (West Conshohocken, PA: ASTM International).
14. Byrne, G., R. Francis, G. Warburton, and M. Maligas, "The Selection of Superduplex Stainless Steel for Oilfield Applications," CORROSION 2004, paper no. 123 (Houston, TX: NACE International, 2004).
15. Byrne, G., "Meaningful Tests for the Quality of Duplex Stainless Steels," CORROSION 2018, paper no. 10876 (Houston, TX: NACE International, 2018).
16. Byrne, G., R. Francis, and G. Warburton, "Variations in Mechanical Properties and Corrosion Resistance of Different Alloys Within the Generic Designation UNS S32760," Duplex America 2000 (Zutphen, Netherlands: KCI Publishing, 2000).
17. ASTM A1084 (latest revision), "Standard Test Method for Detecting Detrimental Phases in Lean Duplex Austenitic/Ferritic Stainless Steels (West Conshohocken, PA: ASTM International).
18. ASTM G28 (latest revision), "Standard Test Methods for Detecting Susceptibility to Intergranular Corrosion in Wrought Nickel-Rich, Chromium Bearing Alloys" (West Conshohocken, PA: ASTM International).
19. Baboian, R., "Corrosion Tests and Standards: Application and Interpretation," ASTM Manual Series MNL 20, 2nd ed., 2005 (West Conshohocken, PA: ASTM International, 2005).

20. Francis R., and S. Hebdon, "The Limiting Thickness for Duplex Stainless Steels," CORROSION 2014, paper no. 3651 (Houston, TX: NACE International, 2014).
21. NORSOK M-650, Qualification of Manufacturers of Special Materials, Rev. 4 (Oslo, Norway: Standards Norway, 2011).
22. ISO 17782 (in development), "Petroleum, Petrochemical and Natural Gas Industries–Qualification of Manufacturers of Special Materials" (Geneva, Switzerland: International Standards Organization).
23. EEMUA Publication 218, "Quality Requirements for the Manufacture and Supply of Duplex Stainless Steels" (London, United Kingdom: Engineering Equipment and Materials Users Association, 2010).
24. Francis, R., G. Warburton, G. Byrne, and J. Wilson, "QA/QC Tools to Ensure the Quality of Duplex Fittings and Other Components," Duplex Stainless Steel 2010 (Zutphen, Netherlands: KCI Publishing, 2010).
25. Francis, R., and S. Hebdon, "The Selection of Stainless Steels for Seawater Pumps," CORROSION 2015, paper no. 5446 (Houston, TX: NACE International, 2015).
26. ASTM A995 (latest revision), "Standard Specification for Castings, Austenitic-Ferritic (Duplex) Stainless Steel for Pressure-Containing Parts" (West Conshohocken, PA: ASTM International).
27. Tossey, B.M., A. Agrawal, and J. Shingledecker, "Effect of Post Weld Clean-up Method on the Localized Corrosion of Duplex Stainless Steels," CORROSION 2014, paper no. 4384 (Houston, TX: NACE International, 2014).
28. Gooch, T., "The Unmixed Zone in CRA Weldments," *Stainless Steel World* 7, 9 (1995): p. 29.
29. NORSOK M-601, Rev 06, "Welding and Inspection of Piping" (Oslo, Norway: Standards Norway, 2015).
30. Talkington, J., D. Harwig, H. Castner, and G. Mitchel, "Development of Titanium Weld Color Inspection Tools," *Welding Journal* (March 2000): pp. 35–38.

Nominal Composition of Some Alloys Commonly Used In Desalination Plants

TABLE A1. Austenitic Cast Irons

ASTM A439 Designation	UNS No.	NOMINAL COMPOSITION (wt%)						
		Fe	C*	Ni	Cr	Si	Mn	P*
D-2	F43000	Bal	3	20	2	2.3	1	0.08
D-2B	F43001	Bal	3	20	3.3	2.3	1	0.08
D-2C	F43002	Bal	2.9	22.5	0.5*	2.5	2.1	0.08
D-2M	F43010	Bal	2.7	22.5	0.2*	2	4.2	0.08
D-3	F43003	Bal	2.6	30	3	2.2	1*	0.08
D-3A	F43004	Bal	2.6	30	1.2	2.2	1*	0.08
D-4	F43005	Bal	2.6	30	5	5.5	1*	0.08
D-5	F43006	Bal	2.4	35	0.1*	2.1	1*	0.08
D-5B	F43007	Bal	2.4	35	2.5	2.1	1*	0.08

Bal = Balance
* = Maximum

TABLE A2. Copper Alloys

TYPE	COMMON NAME	UNS No.	NOMINAL COMPOSITION (wt%)								
			Cu	Zn	Ni	Fe	Mn	Al	Sn	Pb	Other
WROUGHT	Al-Brass	C68700	77	Bal	-	-	-	2	-	-	As
	Muntz Metal	C28000	60	Bal	-	-	-	-	-	0.2	-
	Naval Brass	C46400	61	Bal	-	-	-	-	1	-	-
	DZR Brass	C35330	62	Bal	-	-	-	-	-	2	As
	Mn-Bronze	C67500	58	Bal	-	0.8	0.4	-	0.7	0.1	-
	90/10 Cu-Ni	C70600	Bal	-	10	1.5	1	-	-	-	-
	70/30 Cu-Ni	C71500	Bal	-	30	0.7	0.7	-	-	-	-
	66/30/2/2 Cu-Ni-Fe-Mn	C71640	Bal	-	30	2	2	-	-	-	-
	Nibron	C72400	Bal	-	14.5	1.5	0.3	2	-	-	Mg
	Al-Bronze D	C61400	Bal	-	-	2.5	-	7	-	-	-
	Al-Bronze E	C63000	Bal	-	5	3	-	10	-	-	-
	Nickel Aluminium Bronze	C63200	Bal	-	4.5	4	-	9	-	-	-
CAST	Admiralty Gunmetal	C92400	Bal	2	-	-	-	-	10	2	-
	Phosphor Bronze	C90700	Bal	-	-	-	-	-	10	-	P
	Leaded Bronze	C93700	Bal	-	-	-	-	-	10	10	P
	LG2 Gunmetal	C83600	Bal	5	-	-	-	-	5	5	-
	LG4 Gunmetal	C92410	Bal	2	-	-	-	-	7	3	-
	M-Bronze	C92200	Bal	4	-	-	-	-	6	1.5	-
	Al-Bronze	C95200	Bal	-	-	3	-	9	-	-	-
	Nickel Aluminium Bronze	C95800	Bal	-	4.5	4	-	9	-	-	-

Bal = Balance

TABLE A.3. The nominal compositions of some commonly used wrought stainless steels.

TYPE	UNS No.	GENERIC NAME	NOMINAL COMPOSITION(wt%)								PREN#
			Fe	C*	Cr	Ni	Mo	N	Cu	W	Mn	
Austenitic	S30403	304L	Bal	0.03	18	8	-	-	-	-	1.5	18
	S31603	316L	Bal	0.03	17	10	2	-	-	-	1.5	24
	N08020	Alloy 20	Bal	0.07	20	33	2.5	-	3	-	1.5	28
	N08904	904L	Bal	0.02	20	25	4	-	1.5	-	1	34
Super Austenitic	S31254	6%Mo	Bal	0.02	20	18	6	0.2	0.7	-	0.7	43
	N08367	6%Mo	Bal	0.03	20	25	6	0.2	-	-	1	43
	N08926	6%Mo	Bal	0.02	20	25	6	0.2	1	-	1	43
Ferritic	S41003	410	Bal	0.03	12	0.7	-	-	-	-	1	12
Martensitic	S42000	420	Bal	0.2	13	-	-	-	-	-	0.7	13
Super Ferritic	S44735	Al 29-4C	Bal	0.03	29	0.3	4	-	-	-	0.5	42
	S44660	Seacure	Bal	0.03	26	2.5	3	-	-	-	0.5	36
Lean Duplex	S32304	2304	Bal	0.03	23	4	0.3	0.10	0.3	-	1.5	24
	S32101	2101	Bal	0.03	21	1.5	0.3	0.22	-	-	4.5	26
	S82011	2102	Bal	0.03	21	1.5	0.3	0.22	-	-	2.5	26
	S32202	2202	Bal	0.03	22	2	0.3	0.2	-	-	1.5	26
Duplex	S32205	2205	Ba	0.03	22	5	3.2	0.17	-	-	1	35
	S31803	2205	Bal	0.03	22	5	3	0.13	-	-	1	34
Superduplex	S32760	Z100	Bal	0.03	25	7	3.5	0.25	0.7	0.7	0.7	>41
	S32750	2507	Bal	0.03	25	7	3.6	0.27	-	-	0.8	>41

# PREN = %Cr + 3.3(%Mo + 0.5×%W) + 16×%N
Bal = Balance
* = Maximum

TABLE A.4. The nominal compositions of some commonly used cast stainless steels.

TYPE	ASTM§	GENERIC NAME	NOMINAL COMPOSITION(wt%)								PREN#
			Fe	Cr	Ni	Mo	N	Cu	W	Mn	
Austenitic	CF8M	316	Bal	17	10	2	-	-	-	1	24
	CK3MCuN	6%Mo	Bal	20	18	6	0.2	0.7	-	0.8	43
	CN3Mn	6%Mo	Bal	21	25	6.2	0.2	0.3	-	1	43
Standard Duplex	Grade 4A	2205	Bal	22	5.5	3	0.16	0.5	-	1	35
25%Cr Duplex	Grade 1A¶	CD4MCu	Bal	25	5	2	-	3	-	0.7	32
	Grade 1B	CD4MCuN	Bal	25	5	2	0.14	3	-	0.7	34
Superduplex	Grade 5A	2507	Bal	25	7.5	3.7	0.25	-	-	1	>41
	Grade 6A	Z100	Bal	25	8	3.5	0.25	0.7	0.7	0.7	>41

§ = ASTM designations taken from A351 for austenitic and A995 for duplex
# PREN = %Cr + 3.3(%Mo + 0.5×%W) + 16×%N
Bal = Balance
¶ = withdrawn

TABLE A.5. Nickel Alloys

NAME	UNS No.	NOMINAL COMPOSITION (wt%)									PREN#
		Fe	C	Ni	Cr	Mo	Cu	Al	Ti	W/Nb	
Alloy 400	N04400	2	0.05	65	-	-	Bal	-	-	-	N/A
Alloy K-500	N05500	1.5	0.05	65	-	-	Bal	3	0.6	-	N/A
Alloy 825	N08825	Bal	0.03	40	21	3	2	-	0.9	-	31
Alloy 28	N08028	Bal	0.02	31	27	3	1	-	-	-	37
Alloy 625	N06625	3	0.05	Bal	21	9	-	-	-	3.5Nb	51
Alloy C-276	N10276	5	0.01	Bal	15.5	16	-	-	-	3W	72
Alloy C-22	N06022	3	0.01	Bal	22	13	-	-	-	3W	70
Alloy 59	N06059	1	0.01	Bal	23	15.5	-	0.3	-	-	74
Alloy 2000	N06200	1	0.01	Bal	23	16	1.5	0.25	-	-	76
Alloy 686	N06686	3	0.01	Bal	21	16	-	-	0.1	3.5W	79
Alloy 718	N07718	Bal	0.05	52	19	3	-	0.5	0.9	5Nb	29
Alloy 925	N09925	Bal	0.03	44	21	3	1.8	0.3	2.1	-	31
Alloy 725	N07725	Bal	0.02	56	20	8	-	0.2	1.5	3Nb	46

Bal = Balance
#PREN = %Cr + 3.3(%Mo + 0.5×%W) + 16×%N
N/A = Not Applicable

TABLE A.6. Titanium Alloys

TYPE	GRADE	UNS No.	NOMINAL COMPOSITION (wt%)							
			Ti	O_2*	Mo	Ni	Al	V	Pd	Ru
Alpha	1	R50250	Bal	0.18	-	-	-	-	-	-
	2	R50400	Bal	0.25	-	-	-	-	-	-
	3	R50550	Bal	0.35	-	-	-	-	-	-
	7	R52400	Bal	0.25	-	-	-	-	0.15	-
	11	R52250	Bal	0.18	-	-	-	-	0.15	-
	12	R53400	Bal	0.25	0.3	0.8	-	-	-	-
	16	R52402	Bal	0.25	-	-	-	-	0.05	-
	17	R52252	Bal	0.18	-	-	-	-	0.05	-
	26	R52404	Bal	0.25	-	-	-	-	-	0.1
Alpha-Beta	5	R56400	Bal	0.20	-	-	6	4	-	-

Bal = Balance
* = Maximum

B Ferrous Sulfate Dosing

Ferrous sulfate dosing has been widely used to aid in the formation of protective films on aluminum brass, in particular, but also occasionally with copper-nickel alloys. Ferrous sulfate dosing must be carried out correctly to work properly;[1] failures with ferrous sulfate dosing have occurred when incorrect procedures were used. To help understand why this is important, the role of ferrous sulfate should be understood.

ROLE OF FERROUS SULFATE

When ferrous sulfate is added to water it forms a fine colloid with a negative charge–also called the "zeta potential."[2] With time (tens of seconds) the colloidal particles agglomerate, forming larger particles with a lower charge.[3] The ferrous sulfate works by the colloidal particles being attracted to a metal surface by an extension of the metal's natural electric field, which is stronger in flowing water compared with static water.[2] This explains why ferrous sulfate dosing has little beneficial effect under stagnant conditions, while it has proved to be very efficacious under flowing conditions around 2 m/s, which is typical of that in heat exchangers.

Because the colloid loses its charge as it grows, there is less tendency for it to be attracted to metal surfaces as it progresses through the system. This is why it is recommended that ferrous sulfate be dosed close to, or into, the inlet water box of the heat exchanger being treated.[4] However, where two are in series, experience has shown that injection near the inlet of the first heat exchanger produces a satisfactory film on the tubes in both heat exchangers.

It has also been shown that chlorine accelerates the growth of the colloid and thus reduces its effectiveness, because it is not attracted to metal surfaces as easily.[3] Hence, it is recommended that chlorine dosing to control fouling be turned off while ferrous sulfate dosing is being carried out.[5]

Because the colloid starts growing as soon as the solution is made up, it is also recommended that the solution be made up fresh each time, immediately before dosing into the seawater flow.[3,5]

Because the solution and the seawater are aerated, the ferrous sulfate oxidizes rapidly to an oxy-hydroxide of iron, FeOOH. This is the orange-brown deposit that is found on the tubes, and it works by allowing the natural protective film to form while being shielded from the most aggressive effects of the seawater flow.

DOSING PRACTICE

Because ferrous sulfate decomposes in aerated solution, it should only be made up immediately prior to use. It should be made up in condensate or boiler feed water, because it decomposes quickly in fresh water or seawater. It is usual to make it up as ~10% w/v solution of commercial ferrous sulfate; this should be stirred vigorously until the ferrous sulfate dissolves. This solution has a pH ~3, so care should be exercised when making up the solution.

The solution is dosed into the inlet water box, or as close to it as possible; certainly no more than a few meters upstream. It should be dosed to give a concentration of Fe^{2+} of 1 mg/L for 1 hour. Because chlorine accelerates the colloid growth, all chlorine/chlorine dioxide dosing should be turned off for the dosing time plus 15 min before and after dosing.

With aluminum brass, it has been found that coastal plants require ferrous sulfate dosing once a day for the life of the plant. With copper-nickel alloys, it is often sufficient to dose daily for 60 days until a protective film has formed. Thereafter, the ferrous sulfate dosing can be stopped.

REFERENCES

1. Francis, R., "Survey of the Use of Iron Additions to Seawater Cooled Systems," INCRA Project No. 289 (New York, NY: International Copper Research Association–now the International Copper Association in Washington, DC–January 1979).
2. Gasparini, R., C. Della Rocca, and E. Isannilli, "Mechanism of Protective Film Formation on Cu-Alloy Condenser Tubes with $FeSO_4$ Treatment," *Corrosion Science* 10 (1970): p. 157.
3. Effertz, P.H, and W. Fichte, "Die Eisensulfat-Dosierung in Kuhlsystemen: Grundlagen und Anwendung" ("Ferrous Sulfate Dosing in Cooling Systems: Fundamentals and Application"), *VGB KraftwerksTechnik* 57, 2 (1977): p. 116.
4. Sato, S., "Corrosion and its Prevention in Copper Alloy Condenser Tubes Under Modern Conditions," *Karyoku-Genshiryyoku Hatsuden* 24 (1973): p. 1145.
5. Shone, E.B., and G.C. Grimm, "25 Years Experience with Seawater-Cooled Heat-Transfer Equipment in Shell Fleets," paper no. 11, *Transactions of the Institute of Marine Engineering* 98 (1985).

Index

N

O

T

www.ingramcontent.com/pod-product-compliance
Lightning Source LLC
LaVergne TN
LVHW022211200926
848481LV00006B/1927
9781575903903